THE PURSUIT OF COMPLEXITY

THE PURSUIT OF COMPLEXITY

The utility of biodiversity
from an evolutionary perspective

GERARD JAGERS
2012

Author: Gerard Jagers op Akkerhuis
Edited by: Menno Schilthuizen
English translation/editing: Derek Middleton
Graphic design: lenaleen.nl & welmoet.nl
Fonts: *Auto & Dolly* ©underware.nl
Printing: CPI Koninklijke Wöhrmann

This publication was made possible by contributions of:
het ministerie van EL&I in het Kennisbasis
Onderzoeksprogramma IV van de Wageningen UR
"Duurzame ontwikkeling van de groenblauwe ruimte",
onderdeel Wettelijke Onderzoekstaken Natuur en
Milieu (WOT N&M).

Published by KNNV Publishing
© KNNV Publishing, Zeist, 2012
ISBN: 978 90 5011 443 1
NUR: 738, 740
WWW.KNNVPUBLISHING.NL

This book is a response to a deceptively simple question: Why should we worry about the loss of biodiversity, or, to put it another way, what really is the use of biodiversity? Governments have made conserving biodiversity an important policy objective, but why? Intuitively, the answer seems obvious: the wonder of species richness, preserving the natural heritage for future generations, the useful services and products nature provides, and so on. But the more we think about it, the more complex the question seems to be. For one thing, we do not consider all species to be equally valuable. Besides, how many species are there, and should we give them all equal 'weight'? Are some species more worthy of conservation than others? Or is extinction simply an inevitable part of biological evolution?

The search for an answer to the apparently simple question of the utility of biodiversity forces us to place it within a much broader context than its utility to humans, or even its utility to organisms in general. This book is the outcome of a quest to identify this 'universal utility'.

CONTENTS

1. The utility of biodiversity

Many governmental and international reports, such as the Millennium Ecosystem Assessment (2005), draw a link between biodiversity loss and declining economic prosperity and wellbeing. The value people place on biodiversity is also reflected in the international conventions on its conservation. And because value is generally equated with utility, this book is about the utility of biodiversity.

Why then is biodiversity thought to be so useful? And is the human perspective the most appropriate framework for evaluating and conserving biodiversity? Besides utility to humans, could biodiversity have a more universal value, such as the part it plays in facilitating evolution? In other words, does biodiversity also have evolutionary utility? To answer such questions we first need to clearly define the relevant terms, such as biodiversity and evolution. We also need a framework to guide our thinking about the utility of biodiversity.

So what is biodiversity? The Convention on Biological Diversity (CBD) gives the following definition:

> 'Biological diversity' means the variability among living organisms from all sources including, inter alia, terrestrial, marine and other aquatic ecosystems and the ecological complexes of which they are part: this includes diversity within species, between species and of ecosystems.

Most authors attempt to explain why biodiversity should be conserved by pointing to useful species and the use we make of them. People are worried that useful species will be put at risk if human population growth causes more biodiversity loss. The human species is thus cast in the role of the destroyer of nature. But humans are also just one of all the species on earth, and just like other species we are a product of evolution. If the human species is part of biodiversity, the question of the utility of biodiversity is also about the utility of humans. The concept of utility now assumes a new meaning. It encompasses not only the utility of

biodiversity to humans, but also the utility of humans themselves as a part
of biodiversity.

This chapter begins our exploration by reviewing the most common
arguments for conserving biodiversity. These show that both biodiversity
itself and the question of its utility have many different facets, and that
it is possible to address these topics from many different angles. But
for any discussion to be meaningful it should be clear to everyone what
the concepts and terms used really mean. For this reason, the second
chapter is devoted to deriving clear definitions of the concepts involved,
not as an end in itself, but to give us the clear conceptual tools we need
to develop a reasoned argument. We introduce several important new
concepts, such as the 'acquisition of degrees of freedom' and a general
rule for evolution that treats evolution as if it were a recipe. Just as a recipe
consists of a list of actions and a list of ingredients, evolution also consists
of a series of actions (such as reproduction and selection) and ingredients
(such as parents and offspring). By expanding the two lists of the recipe
for evolution we can devise a very generalised or universal approach to
evolution.

In the subsequent chapters we explore how the processes involved in
the development of the universe led to the origin of life and the evolution
of biodiversity. At the end of the book, both these threads come together
and an evolutionary ladder is used as a theoretical basis for making
predictions about biodiversity in the distant future.

GEOGRAPHICAL DISTRIBUTION OF BIODIVERSITY
AND ECONOMIC PROSPERITY

Many official reports link biodiversity with economic prosperity.
On a global scale, though, there is no clear correlation between the
geographical distribution of biodiversity and economic prosperity.
Although the tropics are richer in species than temperate regions, the

people who live there are often much poorer, but they are poor for many reasons that have nothing to do with the high biodiversity.

First of all, in huge areas like the Amazon basin human poverty or wealth is not related to biodiversity. These areas contain vast numbers of species because of the diversity of habitats they contain, and also because the vast size of the area ensures that a species will almost never go extinct.

Second, a poor economy is not always a result of low biodiversity. The rich Inca civilisation in South America, for example, did not collapse because local biodiversity was lost, but because it actually increased: the Spanish conquistadores brought small pox and other diseases from Europe and as the Incas had no resistance to these diseases they died in large numbers. Recently a convincing case has been made that the collapse of the Easter Island culture, generally viewed as the prime example of the dominant role of ecosystem collapse in the downfall of cultures, was not wiped out by a loss of biodiversity, but because the inhabitants were forced to work in the guano industry and succumbed, like the Incas, to infectious diseases brought from Europe.

Third, the fortuitous occurrence of fossil fuels or valuable minerals also influences the relation between economic prosperity and biodiversity. Valuable minerals can make countries rich, regardless of the state of their biodiversity. Wealth and a thriving economy, on the other hand, usually have a negative impact on biodiversity.

Although biodiversity is not directly related to our prosperity and wellbeing, it is useful to people in many different ways. This raises the question of how people actually perceive the contribution biodiversity makes to their wellbeing.

BIODIVERSITY'S CONTRIBUTION TO ECONOMIC PROSPERITY AND HUMAN WELLBEING

For a long time economists have argued that prosperity is based on technological progress and the division of labour. They consider biodiversity to be useful only when it contributes to economic prosperity, which creates the conditions for wellbeing by enabling people to satisfy their needs and desires. After all, people who cannot afford food or medicines seldom have a high level of wellbeing. In his 1943 paper 'A theory of human motivation', Abraham Maslow presented his theory on the relation between prosperity and wellbeing in the form of his pyramid-shaped Hierarchy of Needs, which still serves as a useful guide. At the bottom of the pyramid are the basic biological and physiological needs, such as food and water. Above that, in ascending order, are safety needs, love and belonging, esteem and self-actualisation. Biodiversity contributes in various ways to the different levels of Maslow's pyramid, for example by providing food and drink, materials for housing, gifts, decoration for our homes and clothing for our bodies, but wellbeing depends on more than just biodiversity.

We can imagine our wellbeing as a big colour chart. The different colours stand for the people in our lives, our home, the school our children go to, our car, the shops in the village, the football field, our partner, access to internet and all the other things in our lives. But how important is nature in the colour chart of our daily lives? And what role does biodiversity play in the colour that stands for 'nature'? Are we bothered about biodiversity as such, or are we mainly interested in cuddly animals and species that are obviously useful for one reason or another? And what colours in the chart, other than nature, are indirectly affected by biodiversity? By asking these types of questions we can discover how biodiversity contributes to people's wellbeing and happiness. The image of the colour chart also shows that it is difficult to take account of all the different contributions made by all the different species. As a solution to this problem, many studies choose to take a practical approach. They consolidate the contributions made by large groups of individual species into what are called 'ecosystem services'.

ECOSYSTEM SERVICES

Nature provides us with all sorts of free services. These 'ecosystem services' include the cleansing of the air we breathe and the water we drink, the provision of edible and otherwise useful biomass, genes for breeding programmes, raw materials for products and medicines, the suppression of pests and diseases by their natural enemies, as well as cultural services, such as attractive natural settings for leisure and recreation. The contributions made by biodiversity to all these services are often cited as examples of the utility of biodiversity.

However, these ecosystem services are not always reliable. This year may be a very good year for fish, for example, while next year there might be bountiful supplies of fruit. This makes it difficult for us to rely heavily on such services. It is why throughout history people have tried to take control of these natural services. Arable and livestock farming are in fact ecosystem services over which we have wrested control from nature, enabling us to obtain higher yields with less effort. And yet the select group of species that have been domesticated as crops and farm animals make up just a fraction of local biodiversity.

The biodiversity nature has to offer has for centuries been used to obtain or improve domestic animals and crop plants by cross-breeding genes from wild plants and animals. Nowadays, advances in biotechnology make it possible to exchange genes between unrelated species. Moreover, many useful genes, for example for antibiotics, come from soil organisms such as fungi, which occur in such vast numbers that their chances of extinction are very slight indeed.

Ecosystem services are slow to react to biodiversity loss. There is a buffer phase in which the species composition changes and only later does the number of the remaining tolerant species start to decline. As the number of species in the ecosystem is much higher than the number strictly needed to provide specific services, it takes time before the numbers have fallen enough to affect these services. Having these extra species is known as redundancy. The fact that fewer species does not necessarily mean lower yields or returns is shown by intensively farmed fields, which produce higher yields but have a much lower biodiversity than surrounding natural areas. Although this high productivity is clearly the result of intensive management, it shows that high biodiversity is not the only yardstick for the proper functioning of ecosystem services. Redundancy means that ecosystem services are not a sufficient means to protect biodiversity. Degraded ecosystems may still provide the ecosystem services we appreciate.

The ecosystem services argument focuses on the utility to humans of natural processes and species, but species redundancy negates any direct relation between biodiversity and ecosystem services. Maintaining ecosystem services is therefore no guarantee that biodiversity will be conserved.

There is also a very different reason why ecosystem services are not a sufficient argument for protecting biodiversity: huge numbers of species provide no service to humans at all, but are rather a source of considerable nuisance.

SPECIES THAT PROVIDE NO SERVICE TO HUMANS

People are choosy about biodiversity. There are many organisms we would rather do without, such as pests and pathogens like mosquitoes, ticks, the bacterium *Borrelia burgdorferi* which causes Lyme disease, rats, termites, lice, hairy caterpillars, maggots, *Phytophthera* pathogens which cause potato blight and other plant diseases, the leprosy bacterium and the malaria parasite. Even magnificent animals like tigers, wolves, bears and elephants would not be welcome in our back gardens.

Our desire to conserve biodiversity is therefore inevitably coloured by our own preferences. People are simply happy to be rid of unpleasant or dangerous things. But other than pests, pathogens and other species that are troublesome in one way or another, people generally regret the loss of species.

THE FEAR OF LOSING SPECIES

Many people think the current rapid loss of biodiversity is a bad thing. The burden put on land and ecosystems by the billions of people on the planet – our ecological footprint – continues to expand and as a result species habitats are degraded or destroyed and their populations decline. When the last individual in a species population dies, the species becomes extinct, and extinction is final. There is no way back. The irreversibility of this process plays a big part in the anxious feelings we have when a species becomes extinct. It means that potential is lost, whereas we prefer to keep as many options as possible open. Moreover, there is the nagging feeling that we ourselves are responsible for the loss of biodiversity.

PERSONAL ARGUMENTS FOR CONSERVATION

The rate of biodiversity loss seems alarming. The fact that humans are agents of extinction makes many people feel uncomfortable and responsible. After all, each one of us could do something to prevent this loss of biodiversity. Besides the practical and economic value of biodiversity, these feelings have their roots in aesthetics, culture, our innate love of the natural world (biophilia) and moral values.

Aesthetics: People can derive considerable pleasure from the beauty and behaviour of plants and animals. We admire colourful orchids, exquisite butterflies, imposing stags, roaring lions, playful dolphins and sociable gorillas. Beauty is an aesthetic argument for wanting to protect these species.

Culture: Our education and upbringing can also influence the value we place on biodiversity. In Western cultures in particular, love for animals is cultivated from an early age. Children are given teddy bears and other cuddly toys to take to bed at night, and they learn to read from picture books in which animals feature prominently. Much broadcasting time, especially in programmes for young children, is devoted to animals, and particularly animals in trouble. For adults there are numerous radio and television programmes about African wildlife, gardening, parks, zoos, flower arranging and numerous other topics related to nature. In contrast, it is interesting to see native children in the South American jungle unashamedly roasting bats over a fire.

This Western love of animals can drive people to extreme acts. Some have been known to leap into a raging river at great personal risk to save their dog – but few would be inclined to do the same for an expensive bicycle. The love of animals even stays with people during early dementia. This may have its origin in our natural preferences, reinforced by influences on behaviour from an early age.

Biophilia: Human evolution may also be partly responsible for our sense of connection to nature. This relationship is known as the biophilia hypothesis. In the same way that the round heads, large eyes and 'charming' movements of babies are highly attractive to adults, our evolutionary selection has made us more favourably disposed to a natural environment most suited to our original ecological niche: a half open landscape containing wooded areas and water. This may explain why people prefer to live in a green environment.

Moral values: Moral arguments tend to be rooted in the belief that the world is a divine creation. Respect for divine works is, in turn, a reason to treat the earth and its inhabitants with respect. On the other hand, religious arguments can be an alibi for exploiting nature. In this view, nature is seen as a gift from God to humanity to use as it sees fit. Based on religious and other convictions, people take different attitudes to nature, ranging (in order of increasing respect) from despot via steward and partner to participant.

People also feel a moral responsibility towards future generations. They will not be able to experience the species that die out now. Will our children miss them? Or our children's children? The Tasmanian wolf and the quagga zebra were exterminated long ago, but what do they mean to us now?

PEOPLE OF THE FUTURE

In *Het Aquarium van Walter Huijsmans* (2009) and *Plastic Panda's* (2011) the Dutch philosopher Bas Haring recently looked for philosophical answers to the question of why we are concerned about the future and why it would be a bad thing if humans cause biodiversity to become impoverished. After all, the species threatened with extinction are already so rare that very few people see them in the wild. Although it may seem odd to want to live in a world without biodiversity, many species could become extinct before the survival of the human race is put in danger.

Haring argues that people's actions are guided largely by their drive to satisfy their current desires and their concern about the sort of lives their children will lead. Haring is interested in whether future generations will be bothered if the natural world contains fewer types of organisms. Of course, it will be problematic if maize or the potato or earthworms are lost, but who blames the seventeenth century sailors for eating all the dodos? The only reminders of the dodo are the reconstructions of what it probably looked like and in cartoons. And what about all the small insect species for which we have no recorded images at all? In fact, the only tangible associations many people have with biodiversity are the plants, ducks and wasps in the park. The biodiversity most people see or know is just a fraction of all biodiversity. This is certainly true for biodiversity far away in a tropical rain forest where people never go. Haring also investigates the origins of diversity and qualifies the importance of species for functional biodiversity, because they seem to be less important than other factors like food supply. Finally, Haring contends that while humans are responsible for a decline in existing biodiversity, they are also a source of a new and different type of biodiversity in the world. This human-related biodiversity includes pedigree dogs, genetically manipulated crops and cultural expressions like fashion clothing lines, architectural styles and plastic pandas. And Haring believes we have every reason to be proud of them.

A FUTURE FOR BIODIVERSITY

It seems as if the significance or importance we attach to biodiversity changes as our living environment becomes increasingly urbanised and technical. In classical antiquity people preferred Arcadian nature to wilderness nature. Arcadia is a form of nature that is 'tamed' to a certain extent by humans; it is a vision of pastoralism and harmony with nature, a pleasant place to be. Wilderness, on the other hand, was wild and threatening. As the industrial revolution gradually turned Arcadian nature into Cultivated nature, people's preferences shifted towards a wilderness type of nature.

This shift in our appreciation of nature suggests that ideas about biodiversity may also change in future. Do people romanticise biodiversity more as they become less directly dependent on it, and will our desire for a biodiverse environment therefore only become greater? Or will the opposite be true and will people in future live happily in big cities surrounded by 'plantations' of solar panels and food or biomass crops?

We cannot know with any certainty what the future will bring, but the general outlines of the future of biodiversity can be predicted. This is the subject of a later chapter. First, though, it is important to explain how we will explore the underlying concepts and the utility of biodiversity.

SCIENTIFIC ARGUMENTS

The biodiversity debate is confused by the jumble of different arguments. Many of these put humans centre stage, almost in opposition to biodiversity, but this approach is internally inconsistent because human beings are also part of biodiversity. An approach is needed in which the human concept of utility or value is itself part of a broader concept of biodiversity and evolution. Such a broad approach requires a context that not only includes human beings, but also takes in the underlying physical and biological mechanisms. In this book all the arguments that incorporate this natural, universal context are called, for want of a better term, scientific arguments. Only when the conceptual background has been brought into focus does it make sense to start finding answers to the more immediate questions, such as what should be conserved, to what degree, where, and at what cost.

Unravelling the concepts and mechanisms underlying the utility of biodiversity is what we call the 'biodiversity question'. Answering this question requires scientific arguments that are consistent with an evolutionary context. After all, the whole of biodiversity, including humans, is a product of evolution. The concept of evolution is interpreted in universal terms in which life on earth is seen as part of the wider evolution of particles and organisms in the universe.

The next question is the important question of how evolution leads to biodiversity. The underlying idea is that biodiversity – including the human species – is not a 'glorious accident'. Biodiversity is indeed glorious, but it is no accident. If the existence of biodiversity is not accidental, we can investigate its origins and vulnerabilities scientifically. Knowing the underlying mechanisms makes it possible to find a better answer to the question of the utility of biodiversity.

The question of the utility of biodiversity to humans and to evolution is a fundamental question, a universal question. It can only be answered by breaking it down into smaller parts, such as:

What is life and what is biodiversity?
What are the causes of biodiversity?
Besides its utility to humans, does biodiversity possess utility from an evolutionary perspective?
What part does the human species play in the evolution of biodiversity?

General answers to these questions will provide a basis for solving practical, social and policy problems, such as:

Why should biodiversity conservation be a policy goal?
What should be conserved, to what degree, and where?
What are the benefits and what are the costs?

The crux of this book lies in laying bare the underlying evolutionary mechanisms and conditions for the emergence and development of biodiversity. This then provides a framework for answering the more practical questions in future studies.

2. Key concepts

Our aim is to investigate the utility of biodiversity. Scientific arguments will be used to explain biodiversity and its utility within the context of evolution, life and the processes that give rise to order in the universe.

Biodiversity is quite a new concept and is much broader than the older terms 'species richness' and 'species diversity'. Species richness is a measure of the number of species living in an area, while species diversity takes account not only of the number of species, but also of the relative abundance of the species in an area.

The word 'biodiversity' is derived from the words bios, life, and diversitas, diversity. Biodiversity therefore simply means the diversity of all life. But what is life, and what factors determine its diversity? How important is evolution for the development of biodiversity, and what exactly is evolution? What does the concept of utility mean when applied to biodiversity?

Sound answers to these and other questions depend on clear definitions of the relevant concepts and terms. Given that life, biodiversity, utility, evolution and similar terms are commonly used in science, it is surprising how little agreement there is about the precise meanings of these terms. This chapter, therefore, is about formulating the clearest possible definitions.

WHAT IS LIFE? (A FIRST STEP)

The issue of what life is comes up several times in this book. On each occasion the answer is further refined using the fruits of the argument at that stage.

The problem of defining what life is has been dragging on for centuries and there is still no generally accepted solution. Some scientists have simply abandoned the search for an answer. They feel the problem of finding a satisfactory definition damages scientific enquiry and diverts attention away from research into the mechanisms that explain life. But how can we know which mechanisms explain life if we do not even know what life is?

It is not the intention here to enter into a long and difficult argument in an attempt to resolve this chicken and egg problem. Instead, we take a pragmatic approach based on the following provisional definition of life:

Only organisms, from simple bacteria to complex animals with brains, meet the definition of life.

This definition is still incomplete, because it does not resolve the problem of circular reasoning: describing organisms as living beings and then defining life by referring back to organisms. This circular reasoning is an intractable problem. In this book we work towards a definition that makes it possible to avoid this problem. We do this in three steps.

The above definition of life states that life is a combination of characteristics that only organisms have in common. This in turn requires a definition of 'organism'. A useful place to start is the following provisional definition:

Individual organisms are descendants of the 'first' cell.

The word 'first' has been put between quotation marks because life probably began with the continual formation of thousands of poorly functioning cells. It is important that this definition does not require an organism to be able to reproduce, but only to be an offspring of its parent or parents. Abandoning the requirement of reproduction solves the problems many definitions have with infertile or otherwise non-reproducing organisms. Here are some examples of these 'exceptions to the rule':

· A sterilised cat cannot reproduce, neither can an old cat.
· Many individuals die before they reach adulthood.
· The offspring of crossings between species are by definition sterile.

In a later chapter we further refine this provisional definition by abandoning the requirement of descent from a parent. This is needed because the first cells cannot have been descendants of an earlier cell, nor is an artificially created bacterium descended from a previous cell.

This approach to defining life has some important consequences for what we do and do not consider to be life. For example, ecosystems are not organisms, because they are not individual descendants of a cell, and so they cannot possess the attribute of life.

WHAT IS DIVERSITY?

Compared with life, diversity is a much simpler concept, because two or more things are diverse if they are different. But because the vast numbers of organisms inhabiting the earth are all unique, the total diversity of all organisms is incalculable. To make the concept of biodiversity workable, organisms have been grouped according to similarity, lineage or habitat, leading to classifications such as 'species' and 'ecosystems'. Such groupings always conceal information about details.

When individual organisms are grouped together into classifications like species, the differences between them, such as age classes, life cycle data and differences between the sexes, are lost. Also unaccounted for are all the different bodily forms of the individuals of the same species resulting from the interaction between gene expression and the environment. These individual forms are called phenotypes. As long as an individual organism is classified as a species, its phenotype remains unknown. We do not know whether an individual is a juvenile or an adult, male or female, thin or a fat, or healthy or sick.

This loss of information cannot be resolved by taking account of genes, because genes do not determine all aspects of the external appearance of an organism. Although genes contain the codes that determine physical form, the phenotype is usually the product of the interplay between gene expression and the physical environment. In fact, focusing only on the genetic part of biodiversity has a surprising consequence: it takes the billions of neural connections in every animal with a complex brain out of the equation entirely. By focusing on phenotypes we explicitly make brain diversity a part of biodiversity.

The ecosystem concept also has its problems. It is difficult to say how many ecosystems there are on earth because all the elements of ecosystems are connected to each other by material flows and by migrations of individuals. In essence, there is really just a single interconnected ecosystem on earth. This ecosystem contains all organisms and also encompasses the abiotic part of the earth that is in contact with, is dependent on or provides a habitat for organisms.

We now turn to the question of defining biodiversity in the light of the above reflections on life and biodiversity. Two well-known definitions from the international literature are given as examples of the types of problems found in many definitions. The first is the definition in the UN Convention on Biological Diversity (1992):

> *'Biological diversity' means the variability among living organisms from all sources including, inter alia, terrestrial, marine and other aquatic ecosystems and the ecological complexes of which they are part: this includes diversity within species, between species and of ecosystems.*

This is the definition cited at the beginning of this book. It refers in general terms to all living organisms, including human beings. Mentioning specific components of the earth's ecosystem and their interconnectedness makes the concept more concrete, but it could be made simpler by saying 'in all places that belong to the earth's ecosystem'. The colon is confusing: it is not clear how what precedes it is explained by what comes after it. If the colon links biodiversity with both the diversity within species (including genes) and the diversity between species and between ecosystems, this implies that genes and ecosystems are part of life. The expressions 'diversity within species' and 'between species' are also ambiguous. Species are taxonomical abstractions and do not relate to parts of an ecosystem.

> *The second definition is by the IUCN (http://www.iucn.org/what/tpas/biodiversity):*
> *Biological diversity – or biodiversity – is a term we use to describe the variety of life on earth. It refers to the wide variety of ecosystems and living organisms: animals, plants, their habitats and their genes.*

In its definition, the IUCN refers first to all life on earth and then cites living organisms, their habitats and their genes. It is not clear why reference is made specifically to living organisms as no-one considers dead organisms (such as petroleum) to be part of biodiversity. The use of the colon in the second sentence again causes difficulties. It is not clear which of the elements before the colon relate to the things listed after it.

Moreover, habitats and genes are not ecosystems, and neither are they organisms.

COMPONENTS OF A DEFINITION OF BIODIVERSITY

What can we learn from the above definitions? First, it appears that the IUCN have deliberately given the concept a broader meaning than the classical concepts of species richness and species diversity. Moreover, they have attempted to reflect different levels of organisation in their definition. With this last objective in mind, they mention the trio of genes, species (or individuals/organisms) and ecosystems. Briefly examining these terms can throw more light on the matter.

Genes

Many definitions take genes to be the yardstick for diversity within species, but focusing on genes ignores a lot of phenotypic diversity. All organisms have genes, but only genes that are uniquely responsible for a single phenotypic characteristic, such as the gene for blue eyes, have a one-to-one relationship with the phenotype. This relationship is often a more complicated one in which the phenotype depends on interactions between multiple genes and between genes and the environment. In such cases genes provide little information about the appearance of the phenotype. Even if we know exactly which genes someone carries, we still know next to nothing about how knowledge changes the structure of the brain, because knowledge is the product of learning experiences. While genes figure prominently in the biodiversity debate, the phenotype – and especially the complexity of the connections within the brain as the carriers of all knowledge – is disregarded.

Moreover, as mentioned earlier, genes are not organisms and are not life. Genes are therefore incongruous elements in a definition of the diversity of life.

Species and individuals

The term 'species' is a taxonomical abstraction that is useful in national legislation and international conventions on the protection of species. The term implicitly refers to all individuals that belong to the population of a species, here called a 'species population'. But in reality, a species population almost always consists of several local subpopulations. A major disadvantage to using the terms species and species population is that neither fit in with ecosystem processes. An individual butterfly or toadstool in an ecosystem knows nothing about categories defined by humans, such as species population. In the natural world the whole species population 'wildebeest' will never fight for its life with the whole species population 'lion', or drink from the same pool. Similarly, a hunter can shoot a fallow deer, but only in exceptional cases can he shoot all the individuals in a species population, and he can never shoot the species fallow deer, because this is just an abstraction. Only individuals ever fight, drink and shoot.

When international conventions talk of protecting a species, they mean that the objective is to ensure that the number of individuals does not fall below the number required to maintain the minimum viable species population. For this reason, in the definitions of biodiversity there is a contradiction between the use of the terms species and species population on the one hand and individuals/organisms on the other. The phrase 'diversity within species' mixes concepts that relate to individuals and to abstractions.

Ecosystems

Ecosystems do not belong in a definition of biodiversity. As the arguments above have clearly shown, the earth's ecosystem does not satisfy the definition of life, despite all the intriguing ideas about Gaia, in which the earth is considered to be a complex self-regulating system. Another approach considers organisms to be walking ecosystems. This idea points to the presence of bacteria within the cells of almost all plants and animals and the presence of small organisms living on the skin and in the intestine, such as bacteria, protozoans and fungi. In this book we consider the bacterial residents that live as symbionts in the body cells of organisms and which have lost the ability to function independently outside the cell (such as chloroplasts in plants) to be part of the organism. In contrast, the small organisms that exist temporarily on or in another organism are, just like the organism they reside on or in, a local part of the earth's ecosystem.

The above reasoning shows that using the terms ecosystem and gene in a definition of biodiversity (the diversity of life!) leads to contrived conceptual links. The phrase 'biodiversity consists of genes, individuals and ecosystems' invites criticism.

THE DEFINITION OF BIODIVERSITY IN THIS BOOK

There are very good reasons to base a definition of biodiversity on organisms. The diversity of organisms covers all phenotypic and genotypic characteristics. Organisms are also the most basic conceptual elements of ecosystems. This makes it possible to construct a simple definition of biodiversity:

> *Biodiversity consists of all the differences between organisms.*

The diversity of organisms is measured at various scales, from global to local. The above definition only mentions the differences between organisms and does not attempt to relate these to measurements of diversity or the conservation of biodiversity, like the genes and ecosystems in the CBD definition. Nevertheless, this new definition does provide a basis for developing conservation strategies. Biodiversity conservation requires the maintenance of all the diversity of organisms and therefore also of all processes upon which this diversity depends. These processes include the interactions between organisms and the interactions between organisms and their environment. All these interactions together define the ecosystem. The ecosystem concept can therefore serve as an umbrella term in the protection of organisms.

The interactions within ecosystems are dynamic: even under natural conditions the abundance of each species varies from year to year. This dynamic nature of ecosystems makes it sensible to aim for a modest conservation objective, for example in the following form:

> *The conservation of biodiversity requires the preservation of a selection of ecosystem elements and their associated processes that in principle guarantee that the numbers of individuals of all species will remain above the minimum required to maintain viable populations, and therefore provide a minimum basis for evolution.*

Moreover, natural evolutionary processes lead to changes in the numbers of species populations in an area, and for this reason biodiversity is not a static concept:

Biodiversity is in a constant state of flux as a result of the evolutionary process in which the number of species populations varies.

The impact of humanity on evolution can itself lead to the creation of new species, for example through genetic manipulation and changes in the environment, which in turn lead to adaptations by species populations. Humans can prevent species extinctions by deliberately protecting certain species and they can cause or hasten species extinctions through their actions. The extinction of species as a result of human action is therefore a natural part of the evolutionary process. So why should people be concerned about the extinction of species? This question leads us right back to the key question of this study: 'Does biodiversity have utility from an evolutionary perspective?'

UTILITY

We have identified two forms of utility: utility to humans and 'universal utility'. We now discuss both forms, starting with utility to humans because we already have a clear picture of what this means. People consider something to be useful if it changes an unsatisfactory situation into a more satisfying one.

Because human beings are organisms with bodies and brains, their needs and desires can result from a mixture of mental and physiological processes. In many cases, mental and physiological needs are inextricably linked. Examples are appetite for food and the desire for sex.

To better understand the word utility, it is informative to ask whether the concept of utility is applicable to organisms with primitive brains, or even no brain at all. Do worms think about utility? Probably not. Do worms experience situations as being more satisfying or less satisfying? We can at least say that through their behavioural mechanisms they do 'try' to avoid unpleasant stimuli. Do worms have desires? Surely: a hungry worm will certainly try to find more food.

If an organism has no nervous system at all, such as a plant, is it still possible to talk of utility? For one thing, plants do not have any mental desires, simply because they have no nervous system. They therefore do not want anything and do not think about utility. But plants do have physiological needs, such as nutrients, water and light, and the growth or cell pressure of a plant can adapt in order to satisfy these needs. If these needs are not met for a long period and phenotypic adjustments do not provide a solution, the plant will eventually die. For people, the purpose of utility is to satisfy needs and desires, including the desire to remain alive; for plants, the purpose of utility is to maintain physiological processes within the limits necessary to ensure their survival. A more general interpretation of the concept of utility therefore encompasses the things or activities that ensure an organism can function normally, or in other words, without too much stress.

But is it also possible to talk about utility with respect to non-living systems, such as molecules and atoms? Or stars? In these cases there is no question of physiological needs; neither is normal functioning a meaningful objective. For these reasons, utility would appear to be applicable only to organisms. All organisms have physiological objectives, while mental objectives are reserved for individuals from a small group of species with brains.

The next chapter shows that the death of organisms can also have utility. Nevertheless, the urge to die as soon as possible seldom delivers any evolutionary advantage to organisms.

The above argument leads to the conclusion that biodiversity cannot be said to have utility for nature because nature is not an organism and therefore does not possess a state of 'normal functioning'.

UNIVERSAL UTILITY

The universe has no known 'normal functioning'. In this sense, the concept of utility has no meaning for the universe. At the same time, all processes and forms in the universe are the result of a single all-encompassing mechanism, which we call 'acquiring degrees of freedom'. Acquiring degrees of freedom leads to greater differentiation in the universe. Differentiation can proceed in two directions: towards more organisation and towards chaos. In both directions nature turns potential into reality. And all the while energy disperses throughout the universe at an increasing rate. What this means precisely requires some clarification.

Acquiring degrees of freedom means the realisation of potentials.

The term 'degrees of freedom' is rather abstract and so it deserves careful explanation. An example of acquiring degrees of freedom is the transition from single-celled to multicellular organisms. A single-celled organism is subject to several limitations. For example, it cannot eat large pieces of food, it cannot grow upwards to where there is more light, and it can easily be eaten by larger organisms. Single-celled organisms that join together can overcome such problems. When acting together, they can eat larger pieces of food, they can grow upwards, and they can become too big to be a prey for some predators. It is because of such advantages that evolution 'invented' multicellular cooperation on numerous separate occasions. Single-celled organisms exchanged their degree of freedom as autonomous entities for the degree of freedom as multicellular entities, gaining access to all the advantages of a larger size and a more complex form.

The concept of degrees of freedom can also be used to describe other forms of integration, for example from an atom (degree of freedom: mono-atomic) to a molecule (degree of freedom: multi-atomic) and from individual nerve cells to brains. Individual bees that together form a colony and individual people in a community also acquire the organisational degree of freedom of collective activity. However, bee colonies and communities are not organisms: earlier on we provisionally defined organisms as individuals that are descended from a first cell, and bee colonies and a communities are not individuals and are not descended from anything.

Not all degrees of freedom are linked to organisation. There are also disorderly, or disorganised, degrees of freedom that are directly connected with the dispersal of energy. Another word for the dispersal of energy is entropy. An important law of nature states that all spontaneous processes lead to an increase in the total entropy of the universe. In simple terms, this means that it takes less energy to make a mess than to clear it up. But like order, clutter is also a degree of freedom.

What does the degree of freedom for clutter look like in practice? This can be explained by examining the death of a multicellular organism. When a multicellular organism dies, it can no longer eat and breathe, and so all the cells in its body slowly suffocate or starve. As the carcass rots the multicellular organism is replaced by a chaotic mass of separate molecules and atoms. The orderly degree of freedom of a living multicellular organism has now been exchanged for the degrees of freedom of separate atoms and molecules.

These examples show that the concept of degrees of freedom can be used to describe processes that lead to order and processes that lead to disorder. It is this capacity to unite changes in both directions that makes the degrees of freedom a crucial concept in this book.

The use of this concept avoids an overemphasis on the increase in disorder and thus avoids the apparent contradictions between the orderly structure of organisms and the law on increasing entropy, or chaos. This contradiction is resolved when we look at nature from the perspective of acquiring degrees of freedom, because nature can acquire orderly (organised) and disorderly (chaotic) degrees of freedom.

The emergence of order in a certain place always involves the dispersal of energy into the surroundings, but because the dispersal of energy is always greater than the corresponding increase in order, on balance, entropy always increases. This universal law also applies to organisms. Growth and metabolism are always accompanied by the degradation of concentrated energy (also called free energy or high-grade energy) from sunlight or from food. Degradation leads to diffuse energy, such as heat and waste, increasing entropy outside the organism. Degraded energy can perform less work. It is less useful. This type of energy is called 'low-grade energy'.

This unbreakable connection between order and entropy leads us to the following conclusion:

> *The entropy organisms create is a necessary consequence of the creation and maintenance of order in (a) their bodies and (b) in the environment (their burrows, communities, cities, etc.).*

Although it is not possible to talk of nature making a useful contribution towards a certain goal, all processes in nature always lead to the acquisition of degrees of freedom. It is therefore possible to say that making a relatively large contribution to acquiring degrees of freedom is always more useful to nature than making a smaller contribution to this process. It is important to clearly distinguish between the concept of utility for organisms and the concept of utility associated with acquiring degrees of freedom. For this reason we call this latter form of utility 'universal utility'.

> *Universal utility is a measure of the relative contribution made by processes to the acquisition of degrees of freedom. Universal utility*

*does not serve a purpose (although it does have direction) and does
not satisfy any needs or desires.*

The universal utility of biodiversity can now be defined as the
contribution it makes to the acquisition of degrees of freedom in the
universe. The greater the contribution made by biodiversity, the greater
its universal utility.

THE BIOLOGICAL DEFINITION OF EVOLUTION

The goal of this book is to investigate the utility of biodiversity from an
evolutionary perspective. To achieve this goal we need clear descriptions
of several fundamental concepts. We have already discussed the concepts
of biodiversity and utility. We turn now to the concept of evolution.

Darwin described evolution as follows. Organisms pass on heritable
and changeable information to their offspring. Each descendant
receives a mix of hereditary factors from its parents, which means
that all the descendants differ in how well they are adapted to a certain
environment. This variation in suitability, or 'fitness', leads to different
capacities for survival.

Darwin talked of hereditary characteristics without knowing
exactly how they are passed on from parent to offspring. And although
Gregor Mendel published his book on inheritance in peas in 1865, it was
some time before this knowledge became incorporated into the idea of
evolution. Genes and DNA were discovered much later. Darwin's ideas
and the science of genetics led to a new synthesis in which evolution
is seen as the change in the relative frequencies of different genes in a
population from generation to generation.

Darwinian evolution cannot explain all evolutionary processes.
For example, viruses also evolve, although they are molecules with a
protein coat, not organisms. Both organisms and viruses multiply by
making copies of themselves, a process called replication. The concept
of replication is broader than that of reproduction and encompasses
a whole range of different ways that things can multiply. The DNA
molecule in a virus *lets* itself be copied. It is a passive process that can
take place in a host cell or in a test tube filled with a nutrient solution
and suitable enzymes. The fact that viruses do not actively copy

themselves makes them very different from cells, which feed themselves and make copies of themselves on their own.

By replacing reproduction with replication it is possible to expand the definition of evolution. But even this does not produce a general concept of evolution. To formulate a truly generic definition of evolution we must first solve another problem. This problem has to do with the similarity between the original and the product of the copying process.

A GENERIC DEFINITION OF EVOLUTION

Making copies never leads to big changes. A copy of a bacterium is always another bacterium. A copy of a bacterial cell can never produce a bacterium in which other bacteria can live (an endosymbiotic cell).

The resident bacteria in endosymbiotic cells take over a lot of the heavy work from the bacterium in which they live, allowing the endosymbiotic cell to become bigger and make more internal membranes. More membranes give the cell the opportunity to neatly package up the DNA inside an additional compartment, the cell nucleus, which protects the DNA against aggressive chemical reactions in the cell. Because endosymbiotic cells generally possess a nucleus they are called eukaryotic cells (*eu* = good and *karyos* = kernel).[1] There are therefore good reasons for the evolution of an endosymbiotic cell, but because endosymbiotic cells are not the same as copies of bacteria, they cannot evolve by making use of a simple copying process.

The copying problem is repeated with the emergence of multicellular organisms. A copy of an endosymbiotic cell never leads to a multicellular organism. A new type of organisation is needed to support multicellularity. Neither are the germ cells produced by multicellular organisms copies of the parent organisms. Although a fertilised egg cell develops into an organism that resembles the parents, the germ cell itself is not a copy, because after fertilisation it still has to go through the whole development process. The copying process in multicellular organisms is a very complex one that involves the development of a fertilised egg cell into an adult phenotype.

---

1   Even though some eukaryotic cells have subsequently lost their endosymbionts while still retaining their nucleus, the emphasis in this book is on the importance of cooperation and so it is appropriate to use the term 'endosymbiotic cell'.

Apparently, major structural changes are key elements in the evolutionary toolkit. They play a role in what is called the symbiogenesis theory formulated at the beginning of the last century by Boris Kozo-Polyanski (*Symbiogenesis*, 1924). This theory emphasises the role of close cooperation between different species during the evolutionary process. The endosymbiont theory, introduced independently many years later by Lynn Margulis (*The Origin of Eukaryotic Cells*, 1970), also seeks to explain the major structural changes that took place through the incorporation of mitochondria or chloroplasts into endosymbiotic cells.

To account for major structural changes and differences in copying processes, we need a definition of evolution that is broader than Darwin's theory, an approach that unites all the above aspects of evolution. It must include an explanation of how original structures (viral DNA, or a cell containing DNA, or a cell containing endosymbionts, or even complete multicellular organisms) give rise to the formation of new, related structures (diversification). The phrase 'give rise to', rather than reproduce, copy or replicate, is expressly chosen to avoid excluding the possibility of cells joining together to make up a new organism, as in the evolution of endosymbionts or multicellular organisms. The functioning of the new structures is then tested in one way or another (selection). The structures that survive this selection process form the starting point for the following round of diversification and selection, and so on. This reasoning leads to a definition of the evolutionary process that goes far beyond the limits of biology:

> The evolution algorithm can be described in a generic way as the repetition of two subprocesses: (1) Diversification, in which an original structure gives rise to the formation of related or derived new structures; and (2) Selection, in which the functioning of the new structures depends on their relative capacities to exist in a certain environment and succeed in diversifying in the next round.

This way of looking at evolution is also discussed by Donald Campbell in a paper published in the journal *Psychological Review* (1960) in which he talks of 'blind variation and selective retention'. Karl Popper, in his book *Objective Knowledge* (1972), also proposed a framework for an alternating process of diversification and selection.

The above definition of the evolution algorithm clearly indicates which processes fall within the concept of evolution and which do not. For example, a process can only be called evolution if it consists of the repetition of the two processes of diversification and selection. During the course of their lives, organisms can age or grow or adapt their phenotype to the surrounding conditions, or even metamorphose, but not evolve. Neither can stars evolve, because stars cannot be said to undergo diversification and selection. Nevertheless, the structure of one star does give rise to the formation of new, related or derived structures: if a star explodes at the end of its life, the gases produced in the explosion eventually come together to form a new star. But it is not possible to identify 'bad' stars which 'drop out' of the process, because there is no context for selection. Ecosystems are equally unable to evolve, because it is not clear how the diversification and selection of ecosystems works or what this process looks like. On the other hand, it is possible to talk of the evolution of insect colonies if the colony depends on the genes of a few key individuals ('queens'). Diversification then operates through the genes of these individuals, because these genes determine the behaviour of the individuals in the colony, and it is the success of the colony compared with other colonies that determines the degree to which the genes of the key individuals participate in the following round of diversification. The division and selection of religious communities and companies also follows an evolutionary process. In this process, diversification occurs through changes in articles of faith and company cultures and selection takes place at the level of the new group.

As long as the repetition of the processes of diversification and selection in the generic definition are met, the evolution algorithm can be applied to a wide range of things, even things that are not organisms. Examples include DNA, 'particles', ideas, cars, computers, computer viruses, insect colonies, companies and many other things.

An example of the evolution of ideas is the evolution of the structure of a car. People repeatedly think up new versions, test them out in practice and continue the process with the best test car. The reactions of car owners are responsible for a co-evolution of our thoughts about cars and the physical realisation of the car as an appliance.

The above generic definition of evolution resembles a recipe, with a list of actions and a list of ingredients. The Darwinian 'list of actions' consists of reproduction, variation and selection. This has now been simplified to the repetition of two successive processes – diversification and selection – which can be interpreted in very broad terms. Darwin's 'list of ingredients' consisted only of organisms and their offspring, but we have now added ideas, cars, computers, computer viruses, insect colonies, businesses and other things to this list. The result is not so much a single general theory of evolution, but rather a framework that indicates for various theories of evolution how general the lists of actions and ingredients are.

With this approach we have made the concept of evolution in principle applicable to all sorts of things and at various levels of abstraction. For example, it can be applied to the evolution of 'particles' from simple types to complex types. The following examples show how this works.

A broad range of conditions in the universe lead to the formation of atoms with different numbers of nuclear particles (diversification). The number of particles in the atomic nucleus and the electron shells around it can take on all the known forms in the isotope map and the periodic table of elements. New processes then determine which atoms are able to form molecules (selection).

During the following repetition of these two steps a large number of different molecules are created (diversification), only a few of which, under local conditions, are able to form cells (selection). This leads to the creation of very many different cells (diversification through reproduction), after which only certain cell lines succeed in developing into endosymbiotic cells (selection). The next steps are the evolution of multicellularity and of multicellular organisms with brains.

In all these steps, selection is connected with the capacity to acquire the next organisational degree of freedom. In organisms, selection is based on survival, but in the transition between particle types selection is based on the realisation of a new degree of freedom.

A remarkable consequence of this abstract approach to evolution is that the evolution of organisms becomes part of the evolution of all types of 'particles'.

By taking an abstract view of evolution it is possible to unify the evolution of particle types and organism types.

Moreover, this sequence of diversification and selection steps automatically leads to a process that generates 'good' outcomes. Evolution is able to solve 'problems' associated with acquiring degrees of freedom without any previous knowledge. A general degree of freedom is the dispersal of energy throughout space during all stages of the evolutionary process. Meanwhile, organisational degrees of freedom are connected either with the adoption of new characteristics in organisms or with the acquisition of new organisational degrees of freedom during the transitions from atoms to molecules, bacterial cells, endosymbiotic cells and multicellular organisms. The acquisition of chaotic and organisational degrees of freedom at the same time gives evolution a high degree of universal utility.

The above reasoning shows that it is possible to take a broad view of evolution consistent with a broad view of utility. As most people still associate evolution with biology, it is useful to give the broad interpretation a special name to distinguish it from biological evolution. Proposals have been made to call this general approach to evolution 'universal Darwinism', in a mark of respect to its founding father. However, this name highlights the biological aspects of evolution and does not do justice to the insights and processes added more recently. A more neutral term is considered appropriate here: 'universal evolution'.

3. The source of biodiversity

In this book we explore the conceptual background to the utility of biodiversity. The previous chapters laid the groundwork by discussing different ideas about biodiversity, before looking in detail at the definitions of biodiversity, utility, universal utility, biological evolution and universal evolution. These have given us the theoretical tools to work with, but we have not yet looked at how biodiversity arises in the first place.

To get to grips with the forces that cause biodiversity, it helps to picture the evolution of biodiversity as a river flowing uphill (Calvin 1987). In the same way that a river cannot flow uphill by itself, there must be a driving force behind biodiversity. Two analogies are useful in unravelling the force behind biodiversity:

the waterfall and the wellspring.

THE UTILITY OF A WATERFALL

Standing beside a big waterfall amid the roar of the water is an exhilarating experience. As the wind brushes your forehead and the mist moistens your hair and face, the question of the utility of the waterfall is not something that immediately springs to mind, and yet that is the question that lies at the heart of this book.

Most people would be puzzled by the question of whether a waterfall can be said to have utility. After all, it is not at all clear who or what should obtain any benefit from the falling water. It has no aim or purpose, it does not satisfy a need or desire, and there is no question of normal functioning (there may be for the observer, but not for the waterfall itself). The term 'universal utility' offers a way forward. It is a measure of the contribution made by processes to the acquisition of degrees of freedom. The question is: What is the universal utility of water falling, or in more general terms, what is the utility of the water cycle on earth?

The question 'What is the utility of the water cycle?' can be stated in the following terms: 'Which process performed by the water cycle contributes to the acquisition of degrees of freedom?' To answer this question we must look for the cause of the water cycle. Starting with rivers, the cycle can be traced back first to rainwater, which feeds the rivers. Rainwater is formed by moisture condensing in clouds, which in turn are formed by the evaporation of water. Water evaporates when the water molecules move rapidly. Molecules move rapidly when they are heated up, and on earth everything (almost everything) is heated by the sun. In the final instance, then, the sun causes water cycles and waterfalls. But this still does not explain why the sun drives the water cycle on earth.

Examining these types of 'why questions' is like peeling an onion. Each question reveals another until we finally arrive at the question 'What is the universal utility of the universe?' Luckily, answering that ultimate question is not the purpose of this book.

THE EARLY UNIVERSE

So, does this line of questioning lead to an open end? Not entirely, because there is no question more fundamental than the origin of the universe. And this gives us somewhere to start: what happened after the very beginning of the universe? Although little is known about the details at that time, the very young and tiny universe expanded rapidly, cooling down in the process as its energy spread out over an increasingly large volume. This can be likened to unscrewing the cap on a bottle of fizzy drink. The compressed air in the drink expands and cools because the same amount of heat is spread out over a much larger volume, which causes the water molecules in the neck of the bottle to condense and form tiny droplets. Similarly, when the universe started to cool down, the energy in it condensed to form highly energetic heavy particles of matter, which continuously formed lighter particles. This process continued until all that was left were the lightest possible particles, which are unable to break down any further. Almost all the matter in the universe now consists in its simplest form of light quarks. These quarks condensed to form nuclear particles, atomic nuclei and then atoms. This process of condensation releases energy, which spreads out through the universe in the form of heat radiation. A huge universe therefore has utility, because a small universe would remain too hot for stable particles to form.

The universe expanded and cooled so rapidly that after just a brief moment the pressure and temperature became too low for nuclear fusion reactions, and so very few heavy atomic nuclei were formed.

This is why 98 per cent of all the matter in the universe that is not unreactive 'dark' matter consists of hydrogen and helium. If the matter in the universe had been packed together for longer, all the atoms in the universe would have been about as heavy as iron, because the spontaneous process of fusion comes to an end when the atomic nucleus achieves the mass of iron.

Making atoms with a mass heavier than iron requires huge inputs of energy and therefore can only occur in conditions that generate huge amounts of energy, for example during the explosion of a star at the end of its life. At such times enough energy is available, for a limited time, to acquire the organisational degrees of freedom of atomic nuclei heavier than iron. Some of this invested energy is released during the fission of heavy atoms like uranium in nuclear reactors.

As a consequence of the rapid expansion of the universe, a huge amount of potential nuclear-fusion energy is locked up in light atoms. This potential fusion energy, released through the nuclear fusion reactor in the sun, is a key driver behind the emergence of biodiversity.

MATTER AND ENERGY

The small particles of matter created early in the history of the universe, such as quarks, nuclear particles and atoms, were distributed evenly across the young universe. This state of affairs did not last for long because all matter creates a gravitational force,[2] which pulls all particles towards each other. In some places, large numbers of particles came together, eventually creating gas clouds of rapidly moving, and therefore hot, particles. These clouds gradually coalesced and cooled by radiating heat into the surrounding universe. They were eventually drawn into denser, galaxy-like formations in which stars, and later planets, were formed as matter continued to coalesce. In stars of sufficient size and pressure, like the sun, the particles in the core are packed together so tightly and at such high temperatures that nuclear fusion reactions take place, creating increasingly heavy atoms. These reactions release the gigantic amounts of energy bound up in hydrogen and helium during the initial expansion of the universe.

---

2   Particles of matter cluster together, making small 'depressions' in space towards which other matter flows.

The sun shines because nuclear fusion releases the potential energy contained in light atoms and radiates this energy as heat into the universe. The earth is heated by the energy radiated from the sun, and this creates the water cycle. This brings us back to the question of the utility of a waterfall.

THE BUILDING BLOCKS OF THE UNIVERSE

What has our search for the origins of biodiversity and complexity in the universe revealed so far? There seem to be two closely interconnected processes at work in the universe. In the early universe, small particles such as quarks, nuclear particles and atoms were formed. Gravity pulls these particles together to form clouds of matter which condense and cluster, taking on various forms, such as galaxies like our own Milky Way, stars and planets. Some of these clusters provide suitable conditions for the emergence of new 'particles', from heavy atoms and simple molecules to organisms. In turn, organisms are responsible for new 'clusters' in the form of ecosystems.

This way of interpreting the genesis of the universe, also referred to as 'Big History', reveals the existence of a process of clustering in parallel with the formation of successive types of particles. New types of particles lead to new types of clusters, and vice versa. These particles and clusters have very different structures.

The clusters display crude forms of organisation, but they do not form successive stages of complexity. Stars cannot create new types of celestial bodies by interacting with other stars. Neither do stars form galaxies; in fact, stars are formed inside galaxies as particles of matter clump together. Particles, on the other hand, display very precise forms of organisation. Particles from the previous level always form the building blocks of larger, more complex particles at the next level. They build on each other to create a 'particle ladder'. Complex particles and organisms are not created by random clustering, but through structured interaction between particles and organisms from the previous level. All physical particles and organisms can therefore be seen as steps on the particle ladder.

Organisms can also be thought of as particles.

Particles and organisms are always formed from building blocks available at the previous level, which are also particles or organisms. This is why organisms have several important features in common with lifeless particles. A characteristic feature of atoms, for example, is the cyclical interaction, the 'physiology', that holds the nucleus together. The protons and neutrons in the nucleus bind together by exchanging smaller particles called pions. Pion exchange changes protons into neutrons and vice versa. Atomic nuclei cannot exist without this continual change in constitution because free neutrons soon disintegrate. In addition, the positively charged nucleus attracts the negatively charged electrons, which form a cloud around the nucleus, called the electron shell.

In some ways, the cyclical nuclear processes and electron shell resemble a bacterium. A bacterium has a cyclical internal physiology: all the molecules in a bacterium continually make and maintain each other, allowing the bacterium to continue to exist. A bacterium also has a shell: the cell membrane, which acts as an interface for the internal physiology and regulates the flow of materials between the cell and its environment.

There are also similarities between particles and organisms at higher levels of organisation. For example, molecules are built up of 'cooperating' atoms and multicellular organisms are built up of cooperating cells.

Accepting that organisms are also types of particles opens up new avenues of investigation. The evolution of organisms can then be seen as a logical continuation of the evolution of non-living, physical particles. In this broad concept of universal evolution, which as explained earlier involves the repeated process of diversification and selection, particles evolve from fundamental particles via nuclear particles and atoms into molecules, and then from molecules to bacteria and to endosymbiotic cells, multicellular plants and multicellular animals.

THE ORGANISM AS ENERGY VORTEX

The orderly structures of organisms, like those of atoms and molecules, are the outcome of a process of self-organisation. Under certain conditions molecules can come together to create a self-maintaining unit. This maintenance is necessary because the molecules in a cell are continually breaking down and so the cell must continually build new molecules to replace those that are lost.

> *In contrast to physical and chemical particles, organisms can only maintain their structure if there is a continuous influx of free energy and building materials from the environment.*

In this respect, organisms can be likened to the 'bathtub vortex' when water drains through the plughole. The vortex is maintained as long as water keeps flowing through the drain. The same goes for organisms, only instead of water, they need a continuous flow of energy and building materials through the drain. An active organism (i.e. one that is not frozen or desiccated) continuously releases energy from the environment, otherwise it cannot function. We can therefore consider organisms as advanced tools which nature uses to convert high-grade energy into low-grade energy. Energy is called 'high-grade' or 'free' if it is still concentrated and can perform a large amount of work, and is therefore useful. The production of low-grade energy reduces the amount of free energy in the universe and increases the amount of entropy.

The degradation of free energy is easy to see in plants. Plants convert energy from sunlight into biomass (concentrated energy) and a lot of rapidly moving molecules of water vapour (dispersed energy). A natural forest degrades solar energy a little bit better than a plantation and much better than desert vegetation. This can be seen on infrared satellite images. Jungles are shown to be cold (blue), plantations are warm (dark green) and deserts are hot (yellow). These differences in temperature are the result of plant growth, which turns concentrated rays of light into a lot of water vapour. This water vapour rises and cools, which leads to the formation of clouds. Because clouds are cold, the radiation emitted from clouds into space is much cooler than the radiation reflected from hot bare soil. A planet with plant cover degrades much more of the sunlight that falls onto it than a planet with a bare surface.

The first bacteria and algae were very primitive and inefficient degraders of energy, whereas human beings, a product of a long period of evolution, are much more advanced. Our energy use per kilogram stands head and shoulders above other organisms. When making such calculations, it is standard practice to include the production of entropy from appliances such as houses, cars, ships, breweries, blast furnaces, oil refineries, power stations, etc. These can all be counted in because all the appliances and machinery used by people only work because of humans and in the service of humans.

All organisms contribute to the acquisition of (orderly and disorderly) degrees of freedom in the universe. The question now is whether this works better if there are more types of organisms; in other words, more biodiversity.

Jumping straight to the biodiversity question is a big leap, though. A step-by-step approach is much easier to follow. The first step is the formation of the universe, which has just been discussed. The second step is the emergence of organisms, the subject of the next section. The last step is the emergence of various types of organisms.

A FOAMING WATERFALL

This chapter began by asking what the utility of a waterfall is. So far we have seen that a waterfall breaks down an energy gradient. Taking this metaphor a stage further, we see that some of the energy in the falling water is used to create air bubbles in the water, which collect as froth on the surface. This can be likened to the formation of a 'biofoam' consisting of separate cells. Water foam is a by-product of falling water; biofoam is a by-product of falling sunlight. A 'sunfall' converts solar radiation containing much concentrated energy into the orderly structures of organisms and a lot of water vapour.

A WELLSPRING OF BIOFOAM[3]

Biofoam possesses an attribute waterfall foam does not. It can use energy from the sunlight to make more foam. Solar energy powers cell physiology rather like water driving a watermill. In every cell,

---

3   Actually, we should also consider the possibility of a chemical waterfall, because chemical energy was once the driving force behind the formation of the very first cells. However, an account of all the possible origins of the first cells is way beyond the scope of this book. We therefore concentrate on life after the emergence of cells that use sunlight to grow.

sunlight fuels a chemical process that converts materials available in the environment into new molecules and new cells. The sunlight forces the cells to carry out this process. They cannot stop. In effect, sunshine inevitably leads to an increase in numbers – a wellspring of biofoam at the bottom of the sunfall. The wellspring continuously pumps biofoam under high pressure into the ecosystem. This pressure, in combination with competition and selection, drives the river of evolution uphill and automatically leads to increasingly complex life forms.

In nature, water wells up out of the ground in artesian springs where the groundwater is under pressure. This pressure forces the water in the artesian spring upwards. In evolution, the enforced increase in numbers is the pressure that drives a struggle for survival. The organisms that do best are those that during their lifetimes, either on their own or in cooperation with others, can acquire more resources than the others (resources are all the things in an ecosystem that an organism can use). How successful an organism is in appropriating resources can only be judged properly over the long term, because the offspring of some parents appropriate more resources than the offspring of other parents.

BIOFOAM CREATES NEW WATERFALLS

The single-celled organisms that live on sunlight and convert it into biomass are called primary producers. Primary producers store up nutrients in their biomass as they grow and reproduce, making it increasingly difficult for subsequent generations to obtain sufficient nutrients, and so population growth slows down. The degradation of free energy stalls. This situation changes as soon as there are other organisms which can release the energy and nutrients bound up in these cells and thus create the conditions for a subsequent waterfall.

> When converting high-grade energy to low-grade energy, it is useful
> if new cells are created that eat the primary producers, digest them
> and excrete the waste products.

The new single-celled organisms that eat primary producers are called herbivores. As the number of herbivores grows, forming the next wellspring, a reservoir of herbivores is created. Again, utility is created for the next type of cell, which eats herbivores: the carnivores. This process in the biofoam at the base of the sunfall creates successive wellsprings and waterfalls of organisms that eat each other – a food chain. Food chains are linked with materials cycles and bind life and death inseparably together.

FOOD CHAINS

Food is an important cause of competition between organisms, and food chains are important pathways of energy flowing through ecosystems. Food chains efficiently convert the high-grade energy of sunlight into low-grade energy in water vapour (plants give off a lot of water in the form of water vapour during their growth) and excrement. This proceeds in a series of alternating wellsprings and waterfalls, creating more organisms which in turn intensify the process. As organisms evolve and adapt to new situations that provide novel sources of free energy, they spread out through the sea, over the land and into freshwater bodies, and during their lives perform useful decomposition work for nature. They do this everywhere, unless they do not succeed in adapting to unfavourable conditions during the evolutionary process.[4]

In theory, a food chain could consist of just a single species at each level: one type of plant that makes biomass from sunlight, one type of herbivore that eats these plants, one type of carnivore that eats the herbivores, and so on. There are simple reasons why there are always more species at each level. During evolution the structure of organisms changes from generation to generation. Nature cannot continually make exactly the same plant, the same herbivore, the same carnivore, and so on. Moreover, no single plant species will be able to make maximum use of all the sun's energy or all the available nutrients everywhere on the planet, and no single herbivore can make optimum use of all the plant biomass to be found everywhere. Because no single organism can do everything, much of the initial stock of resources will remain unused

---

4   There are different theories about the influence environmental factors have on the functioning of organisms, and therefore the selection pressures on them, such as a shortage of nutrients in water (e.g. the significance of the least available nutrient, as demonstrated by Justus von Liebig), a high concentration of toxins (e.g. the development of toxicology by Paracelsus), too high or too low temperatures (e.g. the dependence of enzymes on temperature described by Eyring), and the degree of competition (e.g. the Lotka–Volterra predator–prey equations).

– and unused resources present an opportunity for the evolution of specialised species.

4. The struggle for existence

4 The struggle for existence

When the sun shines, primary producers have to grow and reproduce. Taking it easy is suicide because those individuals that reproduce too slowly fall behind the competition and die. Every organism must therefore fight to remain alive and must waste no time in producing offspring. In this struggle for life, best use must be made of the available resources, not only during the lifetime of each organism, but also by all the generations of descendants of an organism in competition with the descendants of other organisms. During this struggle for existence, on balance evolution leads to increasingly complex life forms. This chapter explains why this is so.

RUNNING WITH THE RED QUEEN

The number of species on earth is not constant. New species are continually emerging while others die out. Palaeontologists have estimated that about 99 per cent of all multicellular organisms that have ever lived on earth have become extinct. Extinction is the rule rather than the exception. So why do species keep on dying out?

Species populations die out when they cannot adapt quickly enough to changes in the ecosystem. To explain this, Leigh Van Valen proposed the Red Queen's Hypothesis, which owes its name to the Red Queen in Lewis Carroll's *Through the Looking-Glass* (1871). The Red Queen is surprised that Alice does not know you have to keep running to stay in the same place. The Red Queen and Alice running is an analogy for the struggle for life and the survival of the fittest. Each species must continually adapt to remain competitive and keep its place in the ecosystem, in other words to stay in the same place.

Adaptation has always been essential, which explains why the evolutionary lifetime of a species, the length of time it exists on earth, has always been about the same. The Red Queen's Hypothesis takes no account of past adaptations. Whales, for example, may appear to have gone through more adaptations (fish to land mammal to marine mammal) than fishes (fish, fish, fish), but they are no better off as a result. The populations of both species have been able to survive because they have been equally good at adapting to the changing environment. The rate at which a species has to adapt depends entirely on the system in which it lives and the way it lives. Well-known 'living fossils' like horseshoe crabs, crocodiles, tortoises, the ginko tree and the Wollemi pine (*Wollemia nobilis*) apparently all live in a quiet backwater on the chessboard. They have survived for millions of years without having had to change very much. On the other hand, many other species have made only a brief appearance. The Red Queen's Hypothesis therefore gives an average picture.

The hypothesis clearly shows that species populations are products of a certain period of time. Species populations can die out if they are not well adapted to their environment, or they can change, or divide into two species if there is room in the ecosystem. Species populations 'flow' through the ecosystem year after year, flowing fastest along the route of least resistance. In this sense, the Red Queen's Hypothesis takes the waterfall model of the breakdown of free energy in the universe and applies it to species.

THE RED QUEEN AND THE CONSTRUCTAL LAW

The Red Queen's Hypothesis and the concept of evolution are linked by a natural law that determines the direction in which the structure of a changing system develops. This law was formulated by Ardrian Bejan and is called the constructal law. It can be stated as follows:

> *For a finite-size flow system to persist in time, its configuration must change such that the flows through the system meet less and less resistance.*[5]

This law is not about the patterns themselves, but the direction in which the patterns develop or evolve. As long as there is a force driving a flow system, the system will gradually, in small steps, put up less resistance to the flow. A hole in a dike provides a good example. As water is forced through the hole it erodes the edges, making the hole bigger and bigger, allowing the water to flow through more and more easily until eventually the dike collapses.

---

5   Bejan's original definition is: *For a finite-size (flow) system to persist in time (to live), its configuration must evolve such that it provides easier access to the imposed currents that flow through it.* Although Bejan includes 'to live' in his definition, the interpretation in this book is restricted to 'persist'. Similarly, 'evolve' is interpreted here to mean 'change'. One reason for this is that not all flow systems represent life, because we have declared above that only organisms can be considered as life. Another reason is that not all changes can be considered to be evolution, which requires repetition of a two-step process of diversification and selection.

The constructal law gives us a way of combining theories of evolution and material flows. But first it is useful to distinguish between two types of flow systems: organisms and the environment.

In organisms, combining evolution and the constructal law means that in each subsequent generation the flows pass through the organisms more easily. This explains the structure of the lungs, kidneys, blood and lymphatic vessels and the nervous system. The structure of the lungs, for example, has evolved into a configuration that provides the optimum balance between air flow and the capacity to take up oxygen and discharge carbon dioxide. This is because increasing the surface area of the lungs can only be achieved by reducing the size of the alveoli, which increases resistance to the inward and outward flow of air. The constructal law also predicts that all processes, organs and bones evolve to a stage at which they can take the same level of stress. This is only logical. Thin bones in a strong body, for example, will not make a strong (competitive) individual. Crucially, the constructal law not only explains relationships between flows and bodily form, but also applies to metabolic 'flows'. Warm-blooded animals, for example, have evolved a faster metabolic rate than cold-blooded animals. Attempts to raise metabolic rate can also be seen in cold-blooded animals, even in fish, which are surrounded by cold water, an example being the relatively high body temperature in tuna. Certain insects, too, such as bees, have a higher body temperature than the air they fly in. Of course, these adaptations rely on the availability of sufficient resources.

The whole ecosystem, including biodiversity, is also itself a flow system. The flow begins with sunlight and ends with water vapour and poorly digestible waste products, such as humus. In principle, the constructal law predicts that all organisms evolve together from generation to generation to reduce the resistance to the material and energy flows involved in converting sunlight into water vapour and waste products. This reduced resistance can be achieved through an increase in biodiversity in which each species provides a channel for the efficient flow of energy and materials through its own particular niche in the ecosystem. But in some situations a single dominant species can meet the requirements of the constructal law very well. Neither does low resistance to flows through the system mean that organisms waste as much energy as possible. Organisms actually use energy very sparingly, because the availability of resources is limited and there is strong competition for those resources. However, it does pay to be more active than your neighbours if being more active allows you to use resources before your competitors can get them. Competition often results in only the most active individuals surviving, because they are better at obtaining resources – but species that consume their resources too quickly will eventually suffer the consequences. This negative feedback mechanism results in a dynamic optimum involving action and reaction. In connection with this, Stuart Kauffman, the author of *The Origins of Order* (1993), argued that biodiversity is highly dynamic, but not quite chaotic. It develops to a point at which it could be said to be balancing on the edge of chaos.

PEOPLE MUST ALSO RUN

The Red Queen also goads people into running fast. She makes companies compete with each other and spurs people on by subtly but effectively exploiting their desire for gratification. This is easy, because people tend to compare their current situation with the past and with other people: always doing the same things is boring, and people do not want to be worse off than their friends and neighbours. Even when we seem to be better off than our neighbours, adverts exploits these feelings to make us feel anxious and dissatisfied and create new desires, which we then do our best to satisfy by buying the products being promoted. At work, too, success depends on the value put on someone's products or output compared with those of others, which leads to competition between colleagues. All these mechanisms keep people running hard to maintain a constant level of satisfaction. Only those who can control their needs and desires well are able to avoid this rat race.

ON THE BRINK OF CHAOS

The Red Queen's Hypothesis offers an explanation for the extinction of species, but it says nothing about the number of species that die out during a particular period. The fossil record shows that species extinction does not always proceed at the same rate. The total biodiversity on earth remains fairly constant for long periods of time, interspersed with occasional bursts of violent change. There are different explanations for these occasional bursts of extinctions.

Sometimes there is a clear external cause, such as changes in solar activity or a meteorite strike, but more important than these external influences is the competition between individuals.

Competition automatically leads to changes in the rate of species extinction. A simple evolutionary game invented by Dante Chialvo illustrates this well (see Figure 1).

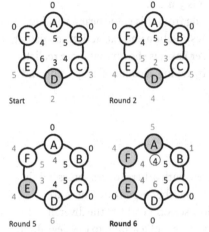

Figure 1. Initial values are shown inside the circle. New values thrown in each round are shown outside the circle. Round numbers printed in bold indicate that the lowest value is equal to or higher than 4. When more than one player throws the lowest value, the circled value is the one that was eliminated by getting the lowest value in subsequent throws.

Six players (species) stand in a circle and each player throws a die, the score determining their level of fitness (adaptedness). The player with the lowest level of fitness becomes extinct (in Figure 1, player D in round 1, who throws a 3) and is replaced by a newly evolved species, with a degree of fitness determined by a throw of the die (the new player D in round 1 throws a 2). If two or more players both throw the same lowest number (as in rounds 3 and 6), the dice are thrown again until one of them has the lowest fitness. Because a new species influences its neighbouring species, its two neighbours in the circle also have to throw again (players C and E, who throw a 3 and a 5). This is the end of the first round. These steps are then repeated in subsequent rounds.

Over many rounds, the average fitness gradually shifts towards a dynamic balance, which fluctuates around a certain 'critical value'. Per Bak (*How*

Nature Works, 1996) says that such critical values occur more frequently in nature, the precipitation of avalanches being one example, and that they lie at a fixed point between the minimum and the maximum. In the case of Chialvo's game the critical value is two-thirds of the maximum, or just under 4. Using this critical fitness value, a wave of extinction can be defined as the number of rounds between two successive moments when the lowest fitness in the game is 4 or higher. In the six rounds in the figure this occurs between rounds 3 and 6.

The waves of extinction resulting from this game bear a striking resemblance to the pattern of species creation and extinction consistent with the Red Queen's Hypothesis. Per Bak has shown that more realistic versions of the game, for example by taking account of adaptation and migration, give the same outcomes.

ARMS RACES AND UTILITY CHAINS

Every organism must always compete with others for resources, while avoiding being used as a resource by another organism. There are therefore two ways of looking at how selection affects the diversity of body forms in nature. The first is a process of learning to use new abiotic resources; the second is an arms race.

As long as various physical and chemical resources are available, organisms benefit from adapting to use them. An adaptation that allows its bearer to use a resource not yet being exploited by competitors clearly delivers a significant advantage. An individual with this adaptation will be able to claim all of this resource available in its location for itself. An environment containing many different resources will therefore usually support many different individuals and species and have a high biodiversity.

These resources are not limited to the physical and chemical (abiotic) environment, but also include the living (biotic) environment: various organisms that are themselves resources for use by other organisms. Evolution therefore produces organisms that are increasingly better at finding and exploiting other organisms. At the same time, organisms develop strategies for not being found and consumed. Why does a hare have long legs? Because there are foxes and birds of prey that can use it as a resource and so it has to be able to run away from them. And why are foxes and birds of prey such clever hunters? Because otherwise they would never catch any hares.

Such bodily adaptations can often be likened to an arms race. Many adaptations enable their bearers to exploit other species as resources, such as big claws, venom, sweet attractants, clinging roots, feather-shaped tentacles and large brains, while others help their bearers to avoid being consumed, such as spines, hair, teeth, poison, good camouflage, the ability to run away very fast, and (again) powerful brains.

To bring order to this enormous diversity of adaptations, it helps to classify them into key types of interaction between organisms and their abiotic and biotic environment. Four main groups are identified here:

1. Interactions based on energy, in which organisms use energy from the environment or are themselves used as an energy resource. Examples are using the energy in sunlight, in the chemical environment and in other organisms.

2. Interactions based on structure, in which organisms make use of the structure of other phenotypes, of forms in the abiotic environment or of materials in food. Examples include orchids that grow on the branches of trees and animals that take up calcium from their food to build bones.

3. Interactions based on information, in which organisms interpret aspects of the environment as signals, or in which they use knowledge or genetic information held by other organisms. Examples include interpreting territorial markings when seeking a partner, the combination of genes during sexual reproduction, and schools for imparting knowledge to children.

4. Interactions based on relocation in space and/or time, in which organisms use other organisms or the flows of wind, water or sand as means of transport. Examples include seeds transported in the gut or on the wool of a sheep to another location. Overwintering in the ground is another way seeds 'time travel' to a future ecosystem.

Separate interactions can also be joined together in chains. For example, a cat that eats a mouse that ate a grain of corn is a food chain consisting of corn–mouse–cat, and if in the jungle a large tree provides a home for bromeliads, which in turn provide a home for mosquito larvae in the pool of water between their leaves, this forms a structure chain consisting of tree–bromeliad–mosquito larva.

Combining the above four types of interaction with the concept of chains provides a simple way of ordering ecosystem processes. The food chain then becomes a combination of interactions based on structure (materials, vitamins) and energy. Some interactions lead not only to chains, but also to loops. Examples are nutrient cycles in which the nutrients flow from the environment through food chains and then back to the environment.

RESOURCE CHAINS AND HUMANS

Human beings have a special place in all the resource chains mentioned in this book.

> *The use of reason, tools and industrial processes, and the power to act with foresight, have made humans so flexible they can use almost all available resources to satisfy their needs and desires.*

Humans are top generalists. We harvest sunlight directly with solar cells and solar power plants, or indirectly from fossil chemical energy stored in coal, oil and natural gas. We grow plants for their fruits, seeds, fibres, rubber, timber and other products. We keep animals for their meat, milk or hide, and to perform work. Individual human intelligence is a resource that can also be used by other individuals. People can get other people to work for them. Wherever and whenever they have the opportunity, humans appropriate free energy from natural energy cascades. We are also increasingly able to replace natural resources with various technically produced substitutes.

> *Human beings are the ultimate substituters.*

These attributes make humans so successful that our numbers have grown to about seven billion. All these people represent a huge source of food, but there are almost no man-eating predators left. Human biomass is now targeted almost exclusively by viruses, fungi, bacteria and other parasites. As almost no-one thinks it is their purpose as part of biodiversity to be a host for viruses or other parasites, the anthropocentric view of biodiversity is biased.

COMPETITION AND COMPLEXITY

Evolution usually leads to complexity – but not always. This can be illustrated by filling a test tube with a rich nutrient solution containing single-celled algae, putting it in a well lit place and continually shaking it gently to distribute the algae evenly throughout the solution. The algae divide about every twenty minutes. If half the mixture in the test tube is replaced with fresh nutrient solution every twenty minutes, all the algae will have a fifty per cent chance of being removed from the test tube. Under these conditions, the algae cannot evolve towards greater complexity; there is simply too little time to construct bigger, more complex algae. Besides, there is no reward for having a more complex structure, because a more complex alga will still have a fifty per cent chance of dying. It is possible, though, for the algae to evolve into smaller, faster algae. If the solution is diluted less frequently, for example every thirty minutes, more and larger algae will be able to grow in the test tube. The conditions now offer opportunities for the evolution of complexity, because there is more time and greater opportunity for competition.

This test tube world shows that evolution does not always lead to complexity. In some situations, organisms die frequently and randomly. In such cases only organisms that can reproduce at a fast rate will survive, and these are mostly the smaller and less complex organisms. In other situations, the chances of dying are low and survival is determined mainly by competition. In these conditions evolution usually leads to greater complexity.

BIG MEETS BIG

Species formation can also be likened to a box of beads. When the box is shaken, the larger beads rise to the top because the smaller beads fall to the bottom more easily. The big beads can only fall down if a large space opens up, whereas the small beads will move down as soon as a smaller space opens up. As a result, the bigger beads rise to the top.

Although shaking a box of beads is a physical process, not a biological one, it illustrates that competition for resources causes the evolution of more complex species. As described above, resources are all those things an organism can use to keep itself functioning. Say that evolution begins with all sorts of small organisms of different sizes in an environment with sufficient resources. The smallest individuals cannot compete so well and therefore obtain fewer of the available resources than the larger individuals, which face competition primarily from the other large individuals. Competition with other large individuals makes it more advantageous for the bigger individuals to evolve towards even larger body sizes. This reasoning applies to all levels and means that evolution through competition for niches leads to the development of large and/or highly competitive species. This process steers the evolution of power, with more power almost always being associated with greater complexity, for example a more advanced structure and/or a larger size. A bird of prey that eats other birds must possess the skills required for a predatory existence. Baleen whales are so big they can eat whole shoals of plankton. Termite colonies developed because individual termites have to be able to crawl through small holes and so they cannot become any bigger; the only way to increase their power was to cooperate in a colony.

A river can only flow upwards under pressure. Likewise, an organism cannot become more complex without the pressure to do so. In nature, this pressure is provided by unbridled reproduction. Competition between offspring then leads to selection of the individuals best suited to a particular situation. Adding organisational degrees of freedom, such as more or better eyes, longer or stronger limbs, or cooperation between individuals in an ant colony, requires energy. Such adaptations cannot survive if they do not deliver a net evolutionary advantage. For example, if having eyesight loses all its competitive advantage, the eyes will degrade, as happened to the blind fish that live in dark caves. Instead of eyes, these fish have evolved a very subtle tactile sense. For the same reason, ant colonies would probably lose their colony structure if there were no longer any competition between colonies. Complexity depends entirely on the continual pressure of unstoppable reproduction combined with competition.

The ecosystem on earth is a dynamic balance between the numbers
of new species populations emerging through evolution and those
becoming extinct. In separate parts of the ecosystem, this balance also
involves immigration and emigration. These two balances together
determine local biodiversity. It helps at this point to imagine an
ecosystem as a pile of sand on a table. The surface area of the table
defines the dimensions of the area and the grains of sand are the
organisms. The emergence of individuals and species populations is
represented by an upward flow of sand from beneath the table through
small holes in the table's surface. The larger the amount of energy and
other abiotic resources in the area, the greater the flow of sand towards
the table's surface. Although at this point the comparison with sand
becomes somewhat tenuous, the distance from the table's surface to
the top of the pile of sand can be seen as a measure of how 'high' an
organism is in the resource chain. For example, a top predator will end
up at the top of the pile of sand. This simple image of an ecosystem can
help us to identify several rules of thumb for biodiversity.

69

The pile of sand rises more quickly and more steeply the more
resources there are in the system. If there are more resources the table
has more 'usable space', which results in longer interaction chains.

Shaking the table hard evens out the pile of sand. The grains of sand
on the table represent organisms that are always looking for a place in
the ecosystem, and thus occupy a position in food chains. When the
table is shaken hard it is impossible for a tall pile to build up, with many
species populations and multi-layered food chains. Biodiversity needs a
certain degree of stability to develop.

The shape of the pile of sand also depends on the properties of the
sand. Smooth round grains will soon roll off the table, whereas damp,
jagged sand grains will stick together to create a tall pile on the table.
The stability of ecosystems depends to a great extent on the interactions
between individuals. If an individual eats just one type of food, or is
highly dependent on a specific resource (a 'round' grain of sand), the
interaction can be said to be 'strong'. Such interactions are susceptible
to disturbance. Individuals that make use of many different resources
(the 'jagged' grains) are more flexible and their larger number of 'weak'
interactions make ecosystems more stable. The properties, or attributes,

of species always influence the interactions between individuals in the ecosystem.

5. Lines of descent and organisational levels

Biodiversity is the outcome of evolution. In turn, evolution is the consequence of the repetition of two processes: diversification and selection. For organisms, evolution means that parents produce offspring with different heritable characters and then the ecosystem selects which of the offspring will take part in the next round of reproduction.

Diversification and selection cause two significant patterns in biodiversity. The first pattern is the division into different species, reflected in what is known as the 'tree of life'. The second pattern is the appearance of the same levels of complexity independently of each other in separate branches of the tree. Examples of these levels are eukaryotic cells and multicellular organisms.

What, though, is the reason for all this variety in nature? How do the same transitions in complexity occur in different branches of the tree? What do the levels of complexity mean for classifying and evaluating biodiversity? And how can organisms maintain stable functioning despite their highly variable genetic material? In this chapter we look for answers to these and other questions about the main patterns we see in evolution.

THE ORIGINS OF GENETICS AND INFORMATION

Cell division, and thus reproduction, is in principle not limited to cells with RNA and DNA. At the very beginning of evolution, cells may have contained only a primitive set of catalytic molecules which, in conjunction with a cell membrane, carried out the reactions necessary to maintain the cell. Cells that made more molecules grew faster and split into two on their own in a form of primitive reproduction. It was probably only much later that a division of labour arose in which long molecules began to carry the information needed to make other molecules which carried out processes in the cell. How RNA and DNA became involved in cell processes is still not known.

The diversity of life on earth is intimately connected with the diversity of 'information' locked in the genetic material in organisms. But RNA and DNA themselves do not represent information, only structure or configuration. The proteins which RNA and DNA code for are also just structures. The actual information they represent is only revealed when we know how DNA contributes to the physiology and structure of an organism. This gives the codes in the DNA a meaning within a context, the condition that has to be met for there to be information. This brief digression on the concept of information lays the groundwork for investigating the acquisition of degrees of freedom of information during evolution.

WHISPERING DOWN THE GENERATIONS

How DNA is a source of biodiversity can be illustrated by the well-known children's whispering game. All the children sit down in a circle. The first child whispers a brief message in the ear of the second child, who passes the whispered message on to the next child, and so on all round the circle. The last child then has to say out loud what he thought the message was. Each correct word receives a point. If there is just one circle, the children can whisper the message as slowly and clearly as possible to get good scores every time. It is more difficult when two teams (two circles) compete and the winning team is the first one to pass on a correct sentence. The children now have to whisper quickly, but also clearly enough in order to pass the message on correctly.

Organisms in nature face the same dilemma. In a stable environment, a cell that develops a control mechanism and copies its DNA very slowly and precisely gains an advantage in a stable environment. But strict control uses up energy and takes more time, which does not help it to compete with its neighbours. This is why cells gain little advantage from having a meticulous control system for their DNA. Moreover, cells with large amounts of DNA will find it increasingly difficult to keep strict control over it. All this makes it counterproductive to make exact copies of its DNA each time. The variability in DNA resulting from copying errors made by cells has been called a biological law by Daniel McShea and Robert Brandon (*Biology's First Law*, 2010). Evolution has to select forms of genetic organisation that enable cells to function stably despite the high changeability of their genes.

For the cell, a little bit of sloppy copying of its DNA appears to be no problem at all. What would happen though if cells were always able to make exact copies of their DNA? If such a cell divides five times it will produce thirty-two offspring with exactly the same DNA. If a virus then appears on the scene, all the offspring will be equally vulnerable to infection. It is therefore not only cheaper, but also strategically advantageous to make an error now and again. Errors create the chance of some offspring having resistance to disease.

Producing offspring with genetic differences has another benefit too. When some of the offspring are resistant, a disease will take longer to find a susceptible individual, which slows down the spread of the disease. In time, as more resistant strains are selected, the search time will eventually become too long for the disease to spread. While small differences in DNA biodiversity protect all resistant individuals against the disease, the susceptible individuals also benefit from the slower rate of spread of the disease. People who refuse vaccination because of their religious convictions benefit from this effect, because the resistance of everyone else indirectly protects them as well. The direct and indirect protection biodiversity gives against disease may be one of the reasons why people sometimes associate biodiversity with ecosystem health.

There has to be a reasonable balance between the occurrence of changes in the DNA and the changes in the environment. In a constant environment, rapid changes in DNA lead to poorly adapted offspring, but an environment that changes too rapidly will outstrip the ability of the DNA to adapt. However, in nature rapid change in environmental conditions is the norm rather than the exception. Organisms have therefore evolved strategies[6] for adapting their genetic material quickly and effectively.[7] Moreover, because it usually takes energy to maintain adaptations (due to the additional genetic information or organs this

6   Of course, species do not 'think' of these adaptations themselves, but this figure of speech makes it easier to understand that it is the individuals with flexible genetic material that tend to survive.

7   Besides mutation, there are various other mechanisms in the cell that can cause changes in the DNA, such as pieces of DNA jumping to another location (transposons), having double helix DNA, putting epigenetic 'clothes pegs' (methyl groups) on parts of the DNA (leaving the structure unchanged, but changing the expression) and histone acetylation.

requires), evolution will quickly weed out such properties when they are no longer needed.

BIODIVERSITY AND INFORMATION

A simple organism needs few genes to code for its physiology, structure and behaviour. A complex organism needs many more genes to do this. Because genes have a coding function, they are often equated with the information needed to keep the cell alive, but this ignores the fact that the information value of genes always depends on their role in the cell: a loose string of DNA is no more than a complicated molecule. What this molecule 'means', what information it carries, only becomes clear in relation to the role played by DNA in the functioning of the cell. This is consistent with the definition by Peter Checkland and Jim Scholes (*Soft Systems Methodology in Action*, 1990): 'Data with attributed meaning in context'; in other words, information equals data plus meaning.

During evolution, nature is constantly looking for new degrees of freedom for the information in organisms. This began in the primitive cell with the coding function of all molecules. Through a process of functional specialisation, the job of transmitting information to the offspring was concentrated in a few large molecules, while the other components of the cell's toolbox took on the tasks of reading off the codes and maintaining the functioning of the cell. The first RNA and DNA molecules in cells already possessed several degrees of freedom: they could continue to exist in a cell, they could change their structure through mutation, and they could relocate with the cell. The degree of freedom for copying conferred the ability to pass on the same 'information' to the offspring. Later, nature acquired the degree of freedom for exchanging information between individuals through sexual reproduction.

Organisms connected to each other over many generations through the exchange of genes create a common store of genes, a gene pool, and are therefore considered to constitute a species population. The species population represents the collective memory of the genes in organisms.[8]

---

8   For every taxonomical species there exists a species population, unless the species is extinct. The species population in turn contains local groups of individuals, which are therefore considered to be subpopulations. If organisms only reproduce asexually, the various lines of descent do not exchange any genes with each other and therefore do not constitute a species population.

For a long period during the history of evolution genes were the only mechanism for transmitting information. This changed with the appearance of organisms with neural networks, which can also store information. In its most primitive form this neural information consists of stored impressions, but it later acquired all the degrees of freedom of genetic information. Moreover, neural information can make use of different types of coding. For example, the image of an egg is linked to the word 'egg', but it can also be linked to the words 'ovum', 'oeuf' or 'ei'. The information about an egg is thus directly linked to comparable terms in other languages and with other conceptual frameworks. Moreover, neural networks can integrate information from different sensory organs. Thinking of a horse can conjure up all sorts of associations, such as the sound of neighing, the clatter of hooves, the smell of leather saddles, the sound of falling horse dung, shiny black riding boots, the training circle at the stables, the taste of smoked horsemeat or other things. Genes are unable to store and alter such multiple associations.

COMPULSORY SEX, RAPID ADAPTATION AND DUMPING 'WASTE'

The view of biodiversity we have built up implies that the two sexes are also a form of biodiversity. Sexual organisms must mate every generation. In principle, two separate sexes are not required: an egg cell could also fuse with another egg cell to create a new organism. The fact that more than one sex is needed is a consequence of the evolution of endosymbiosis.

If two cells containing endosymbionts (such as mitochondria and chloroplasts) merge, the populations of endosymbionts in both cells mix. A power struggle then ensues to decide which population will take its place in the cells of the offspring in the next round of reproduction. As mitochondria and chloroplasts are able to independently reproduce in the cell, a strain that reproduces rapidly will have a competitive advantage over other strains. Conflict between strains affects the efficiency and survival chance of the host cell, so natural selection finds peaceful solutions, such as the rule that just one cell passes on its endosymbionts to the next generation and the other does not. This is the reason for the existence of the male and female forms. In humans, it means that the egg cell always contains a large colony of mitochondria from the mother's side and the sperm cell contains just a few mitochondria, which are immediately killed if they manage to enter the egg cell.

Sexual reproduction involving two sexes is therefore a necessary consequence of the presence of endosymbionts in the cell. But having two sexes is an expensive business because it means that each individual always has to find a partner. In fact, at first sight compulsory sexual reproduction seems to be a counterproductive strategy. An organism that produces sons and daughters which carry only half her genes has a double disadvantage compared with an organism that passes on all her genes to a whole nest of daughters she can produce parthenogenetically (without fertilisation by a male). First, she only passes on half her genes to each of her offspring, and second, she produces sons and daughters which cannot reproduce without the assistance of an individual of the other sex.

However, compulsory sex also has significant advantages. The production of each generation involves the exchange, or recombination, of genes, allowing the gene pool to adapt to changes in the environment more quickly than if all the genes were inseparably bound to each other. Frequent recombination also prevents weak genes remaining in the population. This is because they will regularly come together within an individual and individuals with many bad genes will undergo heavy selection pressure and be unable to pass the genes on to the next generation. This filtering mechanism is not strong enough to entirely eliminate bad genes from a population, but their frequency of occurrence in the population will be kept low.

The advantages of compulsory sex apparently outweigh its disadvantages. The most commonly occurring populations in nature are of sexually reproducing endosymbiotic organisms (both single-celled and multicellular).[9]

---

9 Asexual species sometimes dominate in extreme environments, because the disadvantages of sex start to outweigh the benefits. The asexual species win out because they rapidly produce identical offspring and the extreme levels of stress in effect make the environment 'stable'.

The expressions 'struggle for life' and 'survival of the fittest' imply that organisms possess attributes that give them an advantage over others, such as big teeth, strong muscles, effective camouflage, impressive courtship display and the ability to cooperate. Selection works through the phenotype. It is the phenotype that fights for survival and pays at the cash register of life. A well constructed phenotype is therefore a valuable asset for an organism.

Diversification and selection operate mainly through individual phenotypes, but when individuals cooperate it is the group and not a single individual that has to settle the account at the cash register. When groups compete, the individuals that work well as a group will benefit. Richard Dawkins calls the influence an individual has on other organisms in its environment the 'extended phenotype' of an individual. By cooperating with other individuals, an individual organism may be said to expand its phenotype with what can be thought of as an extra shield consisting of the organisms with which it cooperates. The body of a termite queen may look helpless and fat, but her extended phenotype, in the form of the termite colony, allows her to compete with other colonies. Humans, too, extend their phenotypes by cooperating with other people and surrounding themselves with tools.

BIODIVERSITY IS WHAT IS LEFT OVER

It is now clear that it is difficult to prevent changes in an organism's genetic material. We have also seen that it is actually useful to produce offspring that are not all identical. But that does not mean that every random genetic change adds a new life form to the ecosystem, in the first place because individuals with poor quality genes quickly die (even in the absence of competition) and also because the severe demands made by the environment and competition ensure that the more extreme changes usually have little chance of survival.

A long time ago, during what is called the Cambrian explosion, when nature had just invented primitive nervous systems, a vast number of bizarre life forms appeared and survived. However, this Cambrian fauna did not last long, because the more eccentric phenotypes that initially survived were not able to compete with individuals with more efficient, faster and smarter strategies.

The brief duration of the Cambrian explosion shows that competition and efficiency channel the development of biodiversity along certain 'pathways' or strategies. This phenomenon is called convergence. For example, many animals that eat plankton have sieve-like body parts; all animals that move quickly through air or water have a streamlined form; herbivores have bigger digestive tracts than carnivores; all plants that live independently on sunlight have a large photosynthetic surface area (cacti have a much smaller surface area because of the overriding need to limit evapotranspiration in their arid environments).

The biodiversity on earth is the outcome of a rigorous selection process, which means that almost all variations in body forms represent a direct benefit for the organism in question. Because maintaining useless phenotypic characteristics always uses up energy or has other disadvantages, selection has eradicated numerous useless forms of biodiversity. The biodiversity currently existing on earth therefore represents just a fraction of all the possibilities.

On the other hand, the ability of nature to select out useless structures has its limitations. Throughout nature there are occasional selection backwaters where variations are subject to little selection pressure. Such 'neutral mutations' include having a 'free' or 'fixed' ear lobe and DNA mutations that have no noticeable effects. This background 'noise' does have universal utility, though. Neutral mutations act as a laboratory in which nature experiments with new DNA codes, some of which may provide the foundations for the emergence of new effective mutations.

The power of selection to filter out poorly adapted phenotypes is the main reason why the picture of the descent of all organisms – the tree of life – does not have an endless number of branches.

A TREE OF STRUCTURES

The 'tree of life' shows that evolution leads to a gradual increase in the number of species. The seed sprouts and grows into a tree. Its branches become thicker and produce more and more twigs. The model of a tree is attractive because of its simplicity, but it fails to convey two important features of biodiversity.

The first is that in certain cases, genes can jump from one branch to another like squirrels. This is a source of debate about the structure of the tree, because jumping genes create relations between species on different branches. This jumping behaviour is not limited to individual genes; sometimes whole species, genes and all, manage to jump over to another branch. Certain bacteria and blue-green algae, for example, jumped as endosymbionts (the cell-in-cell form of symbiosis) to a different branch of single-celled organisms. This jumping of whole cells from one branch to another has led to the development of fungi and animals (with mitochondria in their cells) and plants (with mitochondria and chloroplasts in their cells). It is a lesser known fact that nature has carried out many more experiments with the cell-in-cell model, such as the single-celled alga *Peridinium balticum* which contains whole eukaryotic cells as endosymbionts. Recently a snail (*Elysia chlorotica*) was discovered with chloroplasts in its body cells that come from the algae it eats. Such examples show that the branches of the tree of life are much more intertwined than those of a real tree.

The second way in which the tree diagram does not reflect biodiversity is that the 'composition' of the branches changes from bottom to top. The tree of life begins with bacteria. Then certain branches jump over to create endosymbiotic unicellular species, from which some further branches develop with multicellular organisms, some of which produce branches of multicellular organisms that possess neural networks. The tree of life therefore contains several successive transitions in the complexity of organisms and corresponding levels of organisation. These transitions are not related to particular positions in the tree of life. Indeed, the same transitions can usually be seen in many branches of the tree.

The next section explains why it is important to take account of these organisational levels when examining biodiversity. The main reason is that organisms with a certain level of organisation have many characteristics in common. Moreover, the organisational levels help us to build up our knowledge about the structure of organisms step by step, as each new level adds a limited number of entirely new attributes. At the same time, almost all the attributes from the previous levels still come into play at higher levels.

At certain times during the course of evolution the structure of organisms changed dramatically. In their book *The Major Transitions in Evolution* (1995), Eörs Szathmary and John Maynard Smith call these changes 'major evolutionary transitions'. By selecting just a few of these transitions, we can divide organisms into a series of fundamental types of organisation. These are: (1) the bacterial type; (2) the endosymbiotic type; (3) the multicellular endosymbiotic type; (4) the multicellular endosymbiotic type with a neural network.

Each of these types derives its attributes from a totally new form of organisation. These new forms are not just a product of changes in the DNA, nor do they result from having different cell components. Rather, an individual unit arises with a new organisational structure resulting from new connections between units existing at the immediately preceding level.

The major transition from chemistry to biology occurred when a cell was constructed from molecules. To do this, all the molecules involved in constructing the cell had to acquire two important degrees of freedom. To acquire the first degree of freedom, separate catalytic molecules had to come together in a cyclical chemical process in which all the molecules in the cycle are continually manufactured. Acquiring the second degree of freedom involved the construction of the cell membrane. The precise details of the chemistry and the abiotic environment involved in these initial steps remain largely unknown.

The second major transition – from bacteria to the endosymbiotic cell – occurred when some bacteria started living in the cell plasma of other bacteria: endosymbiosis. A cell is an endosymbiotic cell if it is not a parasite and when it can reproduce and divide the offspring between the daughter cells when the host cell divides.

The next major transition was the creation of multicellular organisms. Multicellular organisms consist of cells whose plasma is connected through small pores in the cell walls, although the genetic material cannot move from one cell to another.

During the last major transition in this series, groups of nerve cells began to work together, leading to the development of organisms containing a neural network with sense organs around its edges.

Each successive level is constructed using the 'building blocks' available at the previous level. The attributes created in the lower levels thus work through into the higher levels. Each successive level can be recognised by the appearance of new attributes that define that level. A few examples should clarify this. Genetic material first arose in bacteria and is present in all more complex organisms. The relationship with endosymbionts began in single-celled organisms and is found in all subsequent multicellular eukaryotic organisms. Multicellularity first arose in plants and fungi, but at a higher level it again forms the basis for neural network organisms.

But what is the utility of having different organisational levels? It is that at each new level, organisms acquire several new attributes that give an advantage over the already existing attributes. Proof of such advantages is the repeated independent emergence of endosymbiosis, multicellularity and neural networks in different organism groups. And because the higher levels can do more, it was plants that conquered the land, not bacteria. Bacteria eventually became very successful on land only because multicellular plants composed of eukaryotic cells had first colonised the continents.

Organisational levels not only represent a fundamental ranking of structures in nature, they also provide a frame of reference for evaluating biodiversity.

WHAT IS THE VALUE OF AN ORGANISATIONAL LEVEL?

We have just seen that each successive organisational level depends directly on the structure of the previous level. Without the cell there would be no endosymbiotic cell, without the endosymbiotic cell there would be no multicellular endosymbiotic organisms, and without multicellular organisms there would be no neural-network organisms. As each level is directly dependent on the previous level, the successive levels can be ranked, or numbered: the greater the number of transitions needed to achieve that level of organisation, the higher the number. Working with organisational levels also makes it possible to assess the costs of reaching a certain level. Each successive level has a price that reflects the number of organisms, the number of successive generations and the quantities of resources needed to develop that form of organisation. These are rather like the investment and patents required for an invention. From this perspective, the extinction of organisms with neural networks (animals with brains) represents a much more 'expensive' loss of inventions than the eradication of endosymbiotic cells.

> *By taking account of the resources and inventions needed to achieve subsequent levels, organisational levels provide a framework for estimating how 'bad' it is when a species becomes extinct.*

In view of these organisational levels, it makes sense to take good care of the current organisational level of humanity. The development of human culture has taken very many generations, very many innovative mutations and stupendous amounts of energy. Moreover, it has benefited from the easy availability of resources, such as fossil fuels and mineral ores. Even cultures with relatively little technological expertise were able to use these resources. Our modern technological society developed from primitive cultures which first made use of the most readily available resources and as their knowledge and expertise increased were able to exploit the more difficult reserves. A large and long-term supply of energy is needed, because some inventions, such as computer chips, depend on a huge scale of economic activity.

Human culture is therefore a valuable invention. The question is whether a culture at the level we are now could quickly recover if a catastrophe plunged it back into the Stone Age. It may be that too few resources would be available in sufficient amounts to repeat the development of a technological culture.

6. Future biodiversity

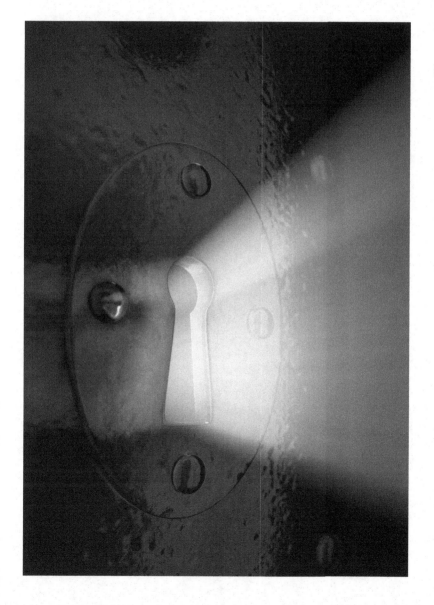

At various times in this book we have referred to this chapter, which gives answers to some difficult questions, including the definition of life and the importance of brain diversity for biodiversity. Another question concerns a second population concept based on the exchange of ideas (alongside the first, which is based on the exchange of genes). Further explanation is also needed for limiting our discussion of the evolution of the complexity of organisms to just four levels. Finally, there is the exciting question of whether this approach can say something about the future of biodiversity on earth.

The most important of these questions for the concept of biodiversity is the definition of life. This is therefore the first question we turn to.

WHAT IS LIFE? THE DIFFERENCE BETWEEN LIFE AND LIVING (STEP TWO)

What is life? is a different question from 'What is living?' Living refers to something that is active, that is 'switched on'. Living is a running leopard, a singing blackbird and the bacteria in rising sourdough bread. But if a plate of sourdough bacteria is put into the fridge and cools down, the process becomes much slower. Our sourdough can even be cooled down to just above absolute zero, at minus 273 degrees Celsius. It can be scientifically proven that at that temperature the atoms in the bacteria cannot move in relation to each other. A bacterium at this extreme temperature is not just slower, but displays no activity whatsoever. Its physiology is turned completely 'off'. Like freezing, total desiccation also brings a bacterium's physiology to a complete standstill. We must therefore conclude that a bacterium in a frozen or desiccated state is no longer alive.

But although it is not alive, could a frozen bacterium still be 'life'? Scientists at the famous Societé de Biologie in Paris asked themselves this question in 1860. Their answer was as simple as it was ingenious: as long as a frozen or desiccated organism can be turned 'on' again it is still life. This conclusion has far-reaching consequences, because it indicates that life is a quality that corresponds to a certain organisational structure of matter.

All activities consistent with 'living' are therefore not relevant to a
definition of 'life'. Breathing, metabolism, perception, reacting to
stimuli, reproduction, evolution and similar activities have no place
in a definition of life.

Putting structure at the core of a definition of life is a way of looking
at life that causes much confusion, because even the most widely used
definitions focus on the characteristics of living. Nevertheless, the
French conclusion is well founded:

If something possesses the material organisation corresponding to life
and it is active, then it is living.

THE MATERIAL ORGANISATION OF LIFE

Understanding that the organisation of matter determines whether
something is life or not is an important step towards a definition of
life. The next step is to find out which type of organisation correlates
with life. Some researchers look for the answer to this in the origin of
life, in the primitive cell. From his experiments into the origin of the
first cell, the chemist Oparin (1924), the pioneer of artificial cellular
life, discovered that it is relatively easy to get spherical bubbles of the
material in cell membranes (phospholipids) to form spontaneously in
water. Years later, the biologists Maturana and Varela (1972) realised that
a definition of life also needs to take account of cell maintenance. Life is
like a clock that winds itself up with energy from the environment, and
even repairs its own cog wheels and hands when they break. Maturana
and Varela called this self-manufacturing capability 'autopoiesis'.
Autopoiesis is a necessary condition for life, because if more material
in the cell beaks down than is made or repaired, the cell itself will
eventually break down.

Unfortunately, knowing that a cell maintains itself does not yet tell
us which structures are required to support this property, but it does
give a clear hint in the right direction. 'Itself' suggests a bounded entity,
while 'maintains' implies a cyclical process. The type of organisation we
are looking for is therefore a spatially defined entity containing a cyclical
process in which the molecules act as a group to maintain each other.

Manfred Eigen and Peter Schuster described a specific form of a cyclical process called a 'hypercycle' (*The Hypercycle: A Principle of Natural Self-Organisation*, 1977-1978). Here we use the term 'hypercycle' to mean a system with a second order process cycle that consists of first order process cycles. An example of a first order process cycle is when an enzyme binds to a substrate molecule, converts it into a product molecule and is then released in the same form as the original enzyme. If there are two enzyme cycles and the product of each cycle is the same as the enzyme in the other cycle, these two enzyme cycles together form an overarching, second order cycle: the hypercycle. In such a hypercycle a group of catalytic molecules can maintain each other by converting substrate molecules from the environment into molecules that make up part of the hypercycle. Besides the hypercycle, a cell membrane is needed. Together, the hypercycle and the membrane ensure that a cell can exist: the hypercycle makes the material for the cell membrane and the cell membrane supports the processes in the hypercycle and holds them together. Without a membrane, the molecules would drift off in all directions and the interactions between them would be lost.

This shows that a material description of life requires interaction between two structures: a chemical hypercycle and an enclosing membrane which is maintained by the internal processes and in turn supports those processes. Structure and function are therefore two sides of the same coin. This definition does not depend on any specific molecules being involved in the processes or the structures, rendering irrelevant any discussion of whether RNA or DNA form the basis of life. Neither does the discovery that bacteria in Mono Lake (America) may use arsenic instead of phosphate in their DNA affect the above approach to defining life in any way. Such questions are only relevant for very specific aspects of life.

The position taken here is that the hypercycle, the cell membrane and the interactions between them provide a sufficient basis for defining the organisation of a bacterial life form. But this does not complete the argument, because not all forms of life are bacteria.

The endosymbiotic cell (the cell-in-cell structure), for example, does not satisfy the above definition. Such cells have at least one other cell living within them, a sort of nested structure with two levels. There is not one cell membrane, but two. There is not one cyclical process, but two. The configuration of an endosymbiotic cell cannot therefore be adequately described by a definition based on a mutual relationship between hypercycle and membrane. The term autopoiesis is not particularly useful here, either, because it recognises no difference between an endosymbiotic cell and a bacterium. Maturana and Varela (1972), who coined the term autopoiesis, were aware of this and proposed a 'second order autopoiesis' to describe the condition of endosymbiosis and multicellularity, and a 'third order autopoiesis' for populations. Below we show that a natural 'ladder' of organisation provides an alternative way to adequately describe these three levels.

The differences between the organisation of a bacterium and an endosymbiotic cell show that the question 'Which configuration of matter can be considered to be the organisation of life?' is not formulated accurately enough. Apparently, we are not dealing with a *single* 'configuration', but rather with *multiple* 'configurations'.

LEVELS OF ORGANISATION

This last realisation opens the way to a more precisely formulated question: Which structural configurations (plural!) must a definition of life include? The answer may be found in the organisational levels introduced earlier in this book: the cell, the endosymbiotic cell, the multicellular organism and the multicellular organism with a neural network. The structure of these levels requires an explanation that invokes a standardised series of specific types of particles and organisms. The levels in this standardised series arise because the configuration of a 'particle' at a lower level is always the direct basis for the construction of a new configuration of a particle at the next level up. It is important at this point to determine whether the interactions give rise to a 'particle' or a system that is itself not a particle, but consists of separate, cooperating particles.

HOW CAN PARTICLES AND ORGANISMS BE RECOGNISED?

The concept of biodiversity is about the diversity of organisms, but how can organisms be recognised among all the other different sorts of systems in nature? Some people say that all things in the universe consist of interactions between smaller components and that therefore all things in the universe are systems. The Milky Way consists of stars and planets, the sun consists of molecules and atoms, and a cow consists of molecules and atoms. The Milky Way, the sun and the cow are therefore all systems. Although this assertion is true, identifying systems with a particle-like nature from all the other systems requires making a distinction. We need to make this distinction because a precise ranking of the complexity of all particles gives us a better understanding of how to define organisms as particles. We also need a clear definition of organisms because they are fundamental to both the definition of life and the definition of biodiversity.

Ever since complexity arose at the beginning of the universe, there have been two ways to create new forms of complexity: either an existing system of particles provides the right conditions for the particles to interact, or such interactions may, in special cases, create the next new particle. A system with these new particles can then give rise to new systems resulting from the interactions between the new particles, or, in certain cases, these interactions create the next new particle. Repetition of this two-stage process generates new systems of interacting particles, which in turn create new particles. In contrast to other systems, the structure of a more complex particle critically depends on the type of particles at the immediately preceding level.

Besides these two ways of creating complexity, there is a third one, which concerns the complexity within a particle. It includes things like a small or large atomic nucleus, organelles in the cell and organs in multicellular organisms.

The above examples show that nature has access to three fundamental degrees of freedom to allow complexity to emerge. Taking the atom as an example of a particle, these degrees of freedom go in the following directions:

· *Inward*: organisation within the atom: the 'internal organisation' of a particle
· *Outward*: interactions between atoms leading to the creation of bigger systems (such as plasma): 'interaction systems'
· *Upward*: the pathway from the atom to the next more complex particle, the molecule

Complex particles cannot simply appear out of nowhere; they are always formed using the types of structures available at the immediately preceding level. Making the transition to the next level of organisation requires a new form of internal organisation or a new form of interaction between existing particles.

> *The formation of a more complex particle is always accompanied by the formation of a new type of spatial configuration and a new type of process. Here, 'new' means that the new attributes are impossible at the previous level of organisation.*

A few examples illustrate this. An atom can create an electron shell around itself, but it takes two atoms to form the covalent bonding characteristic of molecules, in which two or more atoms share electron pairs. Covalent bonding is out of the realms of possibility for individual atoms.

The formation of endosymbiotic cells is another good example. A bacterium has a cell membrane of its own, but 'guest' cells living inside the host bacterium are needed to form an endosymbiotic cell. Endosymbiosis is out of the realms of possibility for individual bacterial cells.

The next step is multicellularity. An individual cell can attach itself to another cell, but a new individual is created only when two cells attached to each other allow the plasma inside their cells to flow between them through pores in the cell membrane. Plasma connections are out of the realms of possibility for individual cells.

Each time an existing particle forms the direct building block for an entirely new attribute, a new 'particle' is born.

PARTICLES + ORGANISMS = OPERATORS

It is difficult to refer to all these configurations as particles when the range of all such particles includes quarks, hadrons, atoms, molecules, bacterial cells, endosymbiotic cells, multicellular endosymbiotic organisms and multicellular endosymbiotic organisms with a neural network. It makes sense to use a word that encompasses both particles and organisms. The term we proposed is 'operator'. Accordingly, the ladder from quarks to multicellular organisms with neural networks is called the 'operator hierarchy'.

The operator hierarchy gives us a new tool for the evolutionary toolkit. Darwin based his theory of evolution on heritable information, differences between offspring and the selection of individuals best suited to their environment. The work of Mendel and Watson and Crick gave us a better understanding of heredity. Eva Jablonka and Marion Lamb (*Evolution in Four Dimensions*, 2005) added processes that temporarily influence the copying of the genetic code over one or more generations.[10] Boris Kozo-Polyanski and Lynn Margulis added a further tool with the idea that genes alone are not able to explain certain transitions in the organisation of organisms. The operator theory now reveals that the combination of cell structures is a biological example of a much more fundamental constructional principle in nature.

The operator theory says that strict and comparable rules for building with forms apply to all operators, both physical particles and organisms. Throughout the whole operator hierarchy, physical laws on structural configuration place restrictions on the transitions between organisational levels.

The operator theory therefore proposes the hypothesis that, at the abstract level of operator types, physical laws governing structural configuration force evolution to proceed according to a fixed pattern.

Of course, the operator theory does not assert that the exact fan-shaped pattern of evolution is fixed in advance. Random genetic changes and selection in a specific environment ensure that the creation of new species remains a random process. What it does say is that the levels of complexity are not accidental. This is because nature must use the existing simple forms as the basis for building new, more complex forms, and because each time it does so, nature must follow strict design laws to meet the requirements of the next higher level of organisation in the operator hierarchy.

By now it should be clear why we limited our discussion of evolution and the complexity of organisms to just a few organisational levels: these are the levels in the operator hierarchy.

It is now time to show that the operator hierarchy is a practical instrument that makes it easier to find answers to the crucial questions posed in this book: the definition of life, the importance of brains for

---

10 This happens when chemical 'clothes pegs' in the form of methyl groups are attached to the structure of the DNA, which then 'reads' differently. Such 'epigenetic' processes do not alter the DNA itself, but rather the way it can be used.

biodiversity, and the future of biodiversity on earth. The main benefit of the operator hierarchy is that it allows us to define life in a way that avoids a circular argument, and it is crucial to define life precisely because it is the basis for defining biodiversity.

WHAT IS LIFE? (THE THIRD AND LAST STEP)

At the beginning of this book we made the provisional assumption that only organisms meet the definition of life and other systems do not. This was accompanied by the warning that defining life by referring to organisms and defining organisms as 'living beings' was a circular argument. The operator theory now offers a solution that avoids this problem of circular reasoning. It describes how the operators in existence at one level determine the nature of the operators that will arise in the next level. This logic avoids the trap of circular reasoning, because each transition leads to a higher level of complexity. The outcome is a long ladder of successive levels of organisational complexity. Each transition to a higher level is accompanied by the creation of two characteristic cyclical processes, a structural one and a functional one, which we call 'closures'. In a bacterium, for example, the cell membrane is the structural closure and the chemical reactions in the cell that ensure the maintenance of the cell physiology form the functional closure. The organisms on each level of the ladder have their own combination of closures, which we call the 'typical closure' for each level.

We can now use this explanation to formulate a new definition of life. First, we decide that from now on all operators with a complexity at least as great as that of the bacterial cell will be called 'organisms'. If we then consider that in the operator hierarchy the organisms at each level of the ladder possess their own typical closure (the specific combination of functional and structural closures), this gives us an easy way to define life: a system is life if it possesses a typical closure found in organisms. If the typical closure is lost, the organism dies and the system is no longer an operator or life.

All operators at least as complex as the cell are organisms.
Life is a general term for the presence of the typical closures found in organisms.

These definitions make it clear that only organisms possess the structural organisation of matter we call life. We knew this already, of course, but the operator hierarchy breaks free of the circular argument 'life = organisms = life' by making use of the organisational ladder of operators in the definition of organisms and then putting the emphasis on the typical closures.

> The above definition of life also implicitly includes future organisms
> as life forms and is therefore open-ended in the 'upward' direction.

Should an organism lose its typical organisational structure, for example if a bacterium no longer possesses autocatalysis and/or a membrane, the system is no longer an organism. It then no longer possesses the attribute of life at that level. The irreversible loss of its typical organisational structure marks the death of the organism.

These definitions can now be put in the following order:

1. *A general organisational ladder*
2a. *All operators at least as complex as a bacterium are organisms*
2b. *Life, focusing on the presence of the characteristic organisational structure of an organism at a certain level*

We now have a sound justification for the choice made at the beginning of this book to recognise only organisms as living beings. Ecosystems and viruses do not meet these definitions of life and organisms because the organisational structure of an ecosystem does not appear in the organisational ladder of operators. A virus molecule with a protein coat does not possess the attributes of the cell because it is not able to replicate itself. The virus molecule is indeed an operator but at a lower level than the cell. This means that neither ecosystems nor viruses belong in a definition of biodiversity.

An important consequence of this is that only a certain type of *structure* determines whether something is alive, rather than its activity or how it is formed or produced. Focusing on these last aspects is the source of the problems encountered in many existing definitions of life. Some are not based on structure, as in the concept of autopoiesis, or are based on just a single type of structure, such as definitions based on the first cell. Other definitions are based on attributes of living systems, such as respiration, metabolism, reaction to stimuli and such like. We discussed why such attributes are irrelevant for a definition of life in a previous chapter, using the example of a frozen or desiccated bacterium.

Two final topics still await explanation: the importance of brains for biodiversity and the future of biodiversity on earth.

MEMES AND IMITATION

Humans are just one species. Estimates of the total number of species on earth vary widely, usually somewhere between 5 and 10 million. Only about 2 million species have been given a scientific name. There are millions of species of bacteria and just a fraction of them have been described by scientists. We have a more complete picture of the endosymbiotic organisms (with the exception of single-celled organisms). The scientific classification of these species includes about 310,000 species of plants, 100,000 species of fungi and 1.49 million species of animals, of which vertebrate animals account for just a small proportion (about 62,000 species).

The human species may appear insignificant among these vast numbers of species, but as a species that can read and write we are unique. Humans are also one of just a select few species able to imitate and learn from each other, along with dolphins, apes, elephants, sea lions and a few other species. People imitate behaviour and ideas, objects, books and other written texts, science and music, as well as codes of conduct, such as those laid down by religions.

The human tendency to imitate is expressed in the concept of memes, proposed by Richard Dawkins (*The Selfish Gene*, 1976) and later also by Susan Blackmore (*The Meme Machine*, 1999). Memes are units of information that can be imitated. Whereas organisms carry genes within their cells, memes reside in their brains. Genes and memes are in many ways similar. Genes are said to display 'selfish' behaviour because they propagate fastest through mechanisms that benefit themselves, while selection ensures that certain genes are frequently found close to each other, for example in chromosomes. Memes also display selfish behaviour and certain memes come together in coherent conceptual worlds (which Suzan Blackmore calls memeplexes). But memes are more flexible than genes because they can also 'hide' as copies in inanimate objects and patterns of sound, where they can 'survive' for long periods. If a meme has embedded itself in an old text on a parchment scroll it can even survive for thousands of years and 'infect' the thoughts of a reader, many generations later, who has been able to decipher the manuscript.

The question now is whether our behaviours and tools (including texts) are themselves the memes, or whether the memes are the neural configurations for behaviour and ideas. Scientists may eventually develop techniques to scan the structure of our brains and convert the information they contain into a row of codes. This coded brain content could then be imitated like a book, making it a meme. Such brain memes would closely resemble genes; they would contain codes for the complex configuration of the brain, rather like genes code for the complex physiology of the cell.

The exchange of memes means that people are not just members of genetic populations, but also members of 'memetic populations'. Human society plus the internet, books and other information carriers thus form a collective meme pool in the same way that a population is a collective gene pool.

This memetic concept of population seems to closely resemble the phenomenon of culture. In this sense, cultures can be seen as complex networks of certain memes, which give rise to generic forms of behaviour. Differences in behavioural patterns distinguish one group of people from another.

THE BRAIN

All memes were thought up at some time or another. As they are passed on from person to person, they are not easily lost. Meanwhile, people keep on thinking up new ones. Remembering all these memes and inventing useful new ones requires a very good set of brains indeed. So how did humans acquire brains that can contain so much knowledge and understanding?

The answer lies in the way information is stored in the brain. The brain does not process impressions as a series of noughts and ones, but as a pattern of interactions between groups of brain cells which are continually firing off signals. Investigating how these patterns work in detail in multicellular brains is extremely hard, but it is possible to make computer models that imitate the basic mechanisms in the brain for storing information. These modelled neural networks reveal that storing information in the brain begins as a rough outline, which is subsequently refined as more information is added.

This way of storing knowledge is like measuring something in increasing detail, for example the length of a certain stretch of coastline. Making a rough measurement using a kilometre-long ruler will give a rough estimate that ignores most twists and turns. Using a metre-long ruler will incorporate all these twists and turns, making the coastline longer. You could also measure the coast using a millimetre-long ruler, which would take in the shapes of individual shells, pebbles and coarse sand grains. This continual increase in the length of the coastline as the ruler used is shortened is also found in fractal patterns. Fractal means broken, which is highly appropriate because each measurement can be broken down into smaller parts, leading to an even bigger total length. The fractal effect appears to make the coastline longer and longer as ever smaller rulers are used. The coastline itself does not change, but its length depends on the scale at which it is measured.

Information is stored in neural networks in much the same way. The learning process in neural networks initially leads to the formation of rough divisions or categories. For example, people learn the difference between dogs and cats, or where the continents of Africa and Europe are found on the globe. These categories are then refined by adding detail. For example, dogs and cats can be divided into breeds, such as terriers, Saint Bernhards, Siamese cats and the European shorthair. Different associations are also made, such as 'all species of cats found in Europe'. By making further combinations of categories, a relatively small neural network can construct a very big 'inner world'.

The diversity of this inner world can be measured in two ways. The first is by looking at the number of nerve cells, the connections between them and the strengths of all these connections. The second type of measurement takes a functional approach which derives the complexity of the neural network from answers to questions, rather like an IQ test.

Everyone knows from experience that the brain contains a vast range of categories and that new categories and patterns of connections are created in the brain every minute. Compared with genes, ideas in neural networks can evolve incredibly quickly, as long as alternative ideas are subject to a selection process.

The evolutionary potential of brains and memes represents a whole new dimension in the development of biodiversity.

The new biodiversity of the brain is evolving at a rapid rate and even does so within a single generation. Is it possible to see where this rapid development is heading?

Predicting the future of all biodiversity is simply too huge a task to be at all realistic. It is impossible to predict future changes in all species of bacteria, all unicellular endosymbiotic organisms, all multicellular organisms and all multicellular organisms with a nervous system. But this is not really relevant. The bacteria alive today are still evolving, but will almost certainly remain bacteria. Even if a new type of endosymbiotic cell does evolve, it will be just one of the very many already in existence. The same reasoning goes for all other species groups. That is why the future depends primarily on the evolution of a new type of structural configuration that is more complex than the neural networks in humans and animals.

FUTURE BIODIVERSITY

The operator theory is a completely new step in the science of evolution and it is still being worked out in more detail. In the meantime, the operator hierarchy offers wholly new avenues for investigation.

> *The operator hierarchy provides a sound basis for extrapolation. It is the first method for predicting the evolution of new structures.*

We can use the operator hierarchy to predict future operators above the level of humans. When making such predictions it is wise to minimise the risk of speculation and so our prediction is limited to the next level above organisms with neural networks. Figure 2 shows that operators at this next level must have the attribute 'structural copying of information'. This type of operator shares this attribute with the cell, which is located in the box under the heading 'predictions'. The cell copies all the information it needs for its functioning by simply copying the structures of all the molecules. This produces a new cell with similar characters to the old one.

> *Because the next type of neural network organism lies in the same column of the operator hierarchy as the cell, it must be capable of*

*systematically copying the structure underlying all the information
contained in the brain.*

Structural copying is not the same as upbringing and learning; it
involves the transfer of information and knowledge simply by copying
the structure of the neural network.

Humans and other organisms with powerful brains do not possess
the ability to read off their neural structures in detail themselves. An
organism with a programmed brain, on the other hand, could very
simply make a copy of the information it contains because the entire
state of the programmed neural network would be stored in files
containing codes for all the separate nerves, the connections between
the nerves and the strengths of these connections. If such files could
be copied directly to a subsequent phenotype, parents would be able
to pass on their knowledge to their offspring without the need for any
upbringing. These 'children' would immediately have the same brains
and knowledge as their parents.

*Assuming that 'structural copying of information' is a required
attribute of the next operator, this leads to the conclusion that the
next operator will inevitably be a technical life form.*

This is an important prediction, because it expands the meaning of
biodiversity. Biodiversity currently includes bacteria, endosymbiotic
cells, multicellular endosymbiotic organisms and multicellular
endosymbiotic organisms with a nervous system. Because they satisfy
the third and last definition of life in this book, technical organisms
– organisms with programmed neural networks – are also part of
biodiversity. It does not matter that they have an engineered or 'artificial'
nervous system.

This last conclusion has surprising consequences for the utility
of biodiversity. If technical organisms are also part of biodiversity, the
common 'organic' basis of biodiversity is eroded, which has several
far-reaching implications. So far, all life forms have been dependent on
physiological processes based on chemical reactions, but this will no
longer be necessary for future 'technical' biodiversity.

Technical organisms will not need plants to produce food, but
will mainly make direct use of the energy in sunlight, for example via
photovoltaic cells. They will probably see plants mainly as a way to turn

solar energy into biofuels or raw materials like plastics. Interactions between technical organisms and non-technical thinking organisms, such as people, will mainly involve the exchange of ideas and the evaluation of certain behaviours.

Technical organisms will be able to colonise and inhabit a wide range of habitats much more easily than humans because they will not be reliant on metabolism to fuel their physiology and will therefore not require an oxygen-rich atmosphere. Technical organisms will be able to put themselves into 'sleep' mode, enabling them to survive for many years without eating, breathing or using power. This will allow them to survive long trips to colonise other planets, building a platform for the evolutionary expansion of technical biodiversity.

Technical organisms will be able to read off their neural structure and store it in a brain code file – a sort of backup of their complete personality – and copy their saved brain code file to another technical body similar to the original. This opens up all sorts of avenues for competition between brain code files for the available technical bodies, analogous to the selfish behaviour of genes in organisms.

It will also be possible to inject 'personalities' into certain members of a group of collaborating individuals in situations where these 'people' can best express themselves. The group will thus be able to use its knowledge to best advantage and generate the most new knowledge. The required stimulus for this behaviour could be competition between groups of technical organisms. The ability to copy brain code files therefore opens up a vast number of utterly unknown pathways for the development of biodiversity.

The above predictions may appear to be a big departure from current biodiversity, and may seem a lot like science fiction. Nevertheless, such predictions are a logical consequence of the operator hierarchy. This glimpse of the future pushes at the boundaries of conventional ideas about biodiversity. If this vision is correct, technical organisms will determine the biodiversity of the future – a biodiversity utterly different from the one we know today.

Figure 2. The evolution of particles and organisms (both 'operators') according to the operator theory. The blue line is the pathway along which operators are formed, a sequence of necessary construction steps. Each of the steps is the shortest route to a subsequent type on the organisational ladder. The systems in the grey columns are theoretical stages and do not have to exist independently.

Definition of abbreviations used in Figure 2:

SAE (Structural Auto Evolution)
An operator with this attribute is able to carry out certain evolutionary processes itself on that part of its structure that contains information.

SCI (Structural Copying of Information)
An operator with this attribute can copy the information contained in its hypercycle by simply copying its structure. This information is contained in the elements of the catalytic hypercycle or in the neural hypercycle.

HMI (Hypercycle Mediating interface)
This type of closure causes an interface with a selective influence on the interactions between the hypercycle and the outside world.

Multi-particle
This type of closure is based in interactions between operators from the level immediately below.

Hypercyle
The type of closure (which stands for a closed process or closed structure) based on cyclical interactions between units that consist of cyclical interactions between units from the immediately preceding level.

Interface
The type of closure that creates a new spatial boundary.

Memon
A collective term for all operators with a neural network.

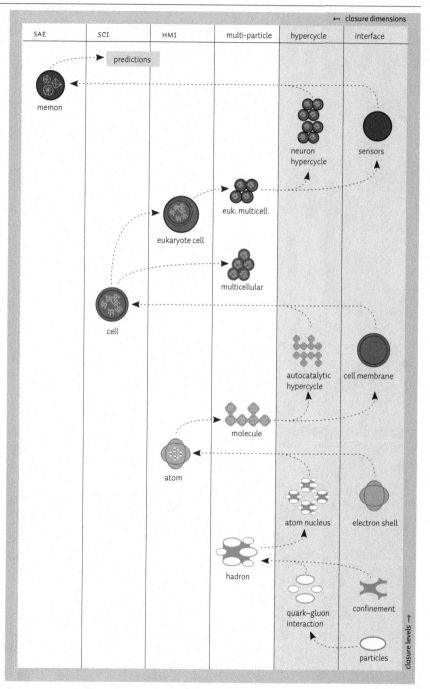

7. The pursuit of complexity

What is the utility of biodiversity? This is the question we set out to answer in our exploration of the evolution of biodiversity in the past, the present and the future. Now, as we approach the end of this book, the time has come to take a step back and assess whether this has brought us any closer to understanding the biodiversity question. This is necessary because the previous chapters do not offer a concrete answer to the question of the utility of biodiversity. But first we briefly review what we have discovered so far.

It is not easy to say what the utility of biodiversity is, for one thing because utility and biodiversity are both difficult concepts. To unravel these concepts, the utility of biodiversity was placed within the broad context of the evolution of the universe. General evolutionary arguments for the utility of biodiversity were called scientific arguments.

To use such scientific arguments, the human species and utility to humans had to be abandoned as the context for the investigation. A new approach to utility was needed, one that applies to biodiversity in general and avoids the artificial distinction between biodiversity and the human species. The focus shifted from 'What is the utility of biodiversity to humans?' to 'What is the utility of biodiversity, including humans, to nature?' This shift in perspective suddenly presented us with a 'universal question' of a much more fundamental nature – intriguing, but much more difficult to answer. To be clear what we were talking about right from the start, the concepts of utility and biodiversity were explored in depth.

The classical concept of utility proved unsuitable for providing a general answer to the question of the utility of biodiversity. Utility is traditionally associated with human judgements about the satisfaction of human needs and desires. While the concept can be used in a more general sense in relation to the needs of other organisms, satisfying needs has no real meaning in relation to the concept of biodiversity. This problem was resolved by taking a new approach based on 'universal utility'.

Universal utility is consistent with the natural law of entropy, which states that all action in the universe occurs at the expense of the concentration of energy. In other words, all processes in the universe dilute energy and raise the level of entropy. To those that equate entropy with disorder, the orderly structures of atoms, molecules and organisms seem to conflict with the laws of entropy. To surmount this contradiction we viewed natural processes in a more neutral light

and analysed them from the perspective of the acquisition of degrees of freedom. Degrees of freedom are all the ways in which energy and matter can be distributed throughout the universe. For example, when the cell emerged, the chemical world acquired the degree of freedom of cellular organisation, and the transition from unicellular to multicellular organisms involved unicellular organisms acquiring the degree of freedom of multicellularity. The term 'acquisition of degrees of freedom' signifies that the corresponding form of organisation always exists in potential; its acquisition involves its material realisation in the universe.

The concept of acquiring degrees of freedom applies to all changes to the state of matter and energy, whether they lead to greater organisation or to chaos. Because the physical laws of conservation always hold true, a local reduction in entropy during the acquisition of orderly degrees of freedom must always be accompanied by a net dispersal of energy in the universe. The concept of universal utility is directly linked to the acquisition of degrees of freedom. Universal utility has nothing to do with a goal, but is simply a 'degree' or 'measure'. So, the more biodiversity contributes to the acquisition of degrees of freedom, the more universal utility it has.

It is easy to make a connection between universal utility and traditional utility to humans. To begin with, acquiring degrees of freedom is a widespread natural process that leads to the development of biodiversity, of which humans are a part. The development of human needs and desires is accompanied by the acquisition of new degrees of freedom (images, wishes) in the brain. Everything that contributes to the satisfaction of their needs and desires has utility to humans.

However, a better understanding and description of the concept of utility is not enough. The concept of biodiversity is also difficult to grasp. The word 'biodiversity' is made up of the terms bios, which means life, and diversitas, which means diversity. Diversity is easy to understand because it simply refers to the existence of differences. Defining life, on the other hand, has always been one of the fundamental questions facing humanity, to which the literature has still not provided a clear answer.

We found a solution to this thorny problem by using the classification of particles and organisms according to the organisational ladder of the operator theory. The operator theory is based on the idea that when nature acquires organisational degrees of freedom, it must follow a series of construction steps defined precisely by laws governing the complexity of structural configuration. All the structures of all types of physical particles and all types of organisms can be ranked according to these steps on a ladder of increasing complexity. This organisational ladder includes the steps from the atom to the molecule, from molecules to the cell and from unicellular organisms to multicellular organisms. All the physical and biological 'particles' in this ranking were given the generic name of 'operator' and the ranking itself we called the 'operator hierarchy'. All assertions about operators and the operator hierarchy constitute the 'operator theory'.

The sequence of operators enables us to identify the levels of organisational complexity and thus provides an independent basis for a definition of life. From this perspective, life is an attribute of the presence of the typical closure possessed by organisms, while the definition of organisms makes use of the particle ladder of the operator theory: all operators at least as complex as a bacterium are called organisms. The operator theory therefore allows independent and fundamental definitions of the concepts of life and organism, avoiding the circular arguments that have caused so many problems in the past.

The definitions of life and organisms now provide a solid platform for defining biodiversity: biodiversity consists of all the differences between organisms. This definition of biodiversity has been purposely kept distinct from the conditions for protecting biodiversity. Many of the definitions of biodiversity currently in use are larded with aspects relating to measuring and/or conserving it, but these aspects must be kept separate if we are to formulate a clear definition of what biodiversity actually is. To conserve a minimum level of biodiversity it is necessary to maintain a selection of ecosystem components and associated processes that will ensure the continuation of differences between organisms. This includes creating the conditions for the maintenance of ecosystems in which the population sizes of species remain above their minimum viable level. This in turn guarantees a platform for the evolution of the species concerned and sustains the possibility for new species to emerge and existing species to die out. In this process, humans are just one species like all the others. Ultimately, conserving biodiversity means conserving the human species.

This brief review shows that the ideas presented in this book overcome several obstacles to clearly defining the key concepts of utility and biodiversity. This preparatory work opens the way to answering the central question in this book: 'What is the utility of biodiversity?' The next and final part of this book gives an answer to this question in three steps. These deal in turn with universal utility, the utility of biodiversity to humans (as part of biodiversity) and the utility of humans (as part of biodiversity) to biodiversity.

THE UTILITY OF BIODIVERSITY

Universal utility is the broadest possible context for talking about the utility of biodiversity. The universal utility of biodiversity is the contribution it makes to the acquisition of degrees of freedom. The acquisition of degrees of freedom drives the development of the earth's ecosystem and the species it contains towards increasingly efficient conversion of energy from the sun into low-grade waste.

Inanimate nature automatically develops efficient flow patterns. For example, a crack in a dike will eventually wear away to such an extent that the dike itself will break. But this natural 'erosion' does not happen automatically in organisms. Evolution is needed to ensure that successive generations of all organisms on earth collectively degrade the sun's energy more efficiently, completely and rapidly. In this process all parts of organisms, such as their physiology, organs and bones, evolve towards comparably favourable levels of efficiency and stress.

At the level of all biodiversity, evolution continually gives rise to new species groups that together offer increasing access to the material and energy flows. All these processes turn sunlight and other energy sources (such as the fossil energy being used today) into water vapour and waste products. Both individual organisms and all of biodiversity, everywhere and always, add to universal utility as much as they can.

A small part of universal utility is the utility of biodiversity to organisms for meeting their needs. As all organisms have evolved in interaction with each other and in competition with each other, they often depend on each other to satisfy their needs. This means that species cannot become extinct without further consequences, but nature itself constantly disrupts the status quo by producing new species. In response to this dynamic, most interactions, and therefore the structure of ecosystems, have necessarily evolved to become more robust.

It is also possible to look at biodiversity in relation to a single species: the human species. The utility of other species then depends on the degree to which they help to satisfy human needs and aspirations. Through our access to technological aids, we have become the top generalist, able to make use of virtually all the components of the earth's ecosystem. The human species has thus effectively reduced its direct dependence on biodiversity. This achievement is the result of several different developments: humans can use their brains to rapidly adapt their behaviour, their knowledge allows them to substitute the services of one species for those of another, and technological aids give humans great control over their environment.

The third and final step focuses on the utility of humans as part of biodiversity. What evolutionary role do humans play? From the perspective of universal utility the productivity of human beings has never been equalled. Human activities increase both chaos and order in the universe at an enormous rate. All the energy flows that are brought together in the service of each kilogram of humanity amount to immensely more than any other animal species can come close to achieving. People extract concentrated stocks of resources and distribute them around the globe. Entropy has never before increased at such a rate.

How is biodiversity reacting to all this? In ecosystems without humans, nature makes use of all 'natural' possibilities for producing entropy. At the same time, the species compositions of ecosystems evolve towards the least possible resistance to the conversion of solar energy into low-grade waste. Each species has its own role and develops to maximise its use of and dominance over resources. Depending on the surrounding conditions, in some places this leads to high biodiversity and in other places it does not. In some locations natural selection favours species adaptations to specific circumstances more than the presence of many species.

Ecosystems containing humans are subject to considerable change, but evolution's pursuit of all available pathways to maximum acquisition of degrees of freedom remains the same. And human beings are not just a pathway, but a veritable highway. Humans have released fossil energy from its 'prison', giving humanity the opportunity to greatly expand its level of activity while increasing the flows in the ecosystem. For example, the highly productive species and cultivation technologies used in intensive farming systems deliver higher yields and create more entropy than found in nature. The original natural systems, irrespective of their biodiversity, are less productive per unit time and are 'overtaken'. Although intensive agricultural and urban systems create more order in the form of agricultural products and culture, these systems also generate much entropy through their use of fuels, fertilisers and other inputs. Both orderly and disorderly degrees of freedom are acquired at a rapid rate, giving intensive systems a relatively high universal utility.

As always, action leads to reaction. Global warming, the energy crisis, resource scarcity, wars, biodiversity loss, shortages of agricultural land, overpopulation and epidemics can be seen as reactions, or brakes, on this process. But species in ecosystems without people are also subject to limiting factors. Evolution resolves such problems by always finding paths towards maximum use of existing possibilities. Since the arrival of the human species on the scene, knowledge development and policy have become a part of evolution. Human efforts in the materials, biological, medical and social fields have long been successful in bypassing major constraints on maximising the degradation of energy and the continued production of order.

Humanity does not produce so much entropy without reason. The consumption of free energy is 'payment' for the order humans create. For one thing, during much of their lives human brains generate order very quickly. People also maintain their bodies, and they design and build roads, houses, cities, agricultural systems, factories, radios, cars, computers and other tools to a high degree of organisation. Eventually, people will use computer simulation of neural networks to make tools that can maintain and increase their organisational autonomy. According to the operator theory, this type of intelligent tool is the next evolutionary step.

The Red Queen drives us on faster than we realise. Humans are also just pawns in *her* game of chess, in which evolution and biodiversity repeatedly create the conditions for new rounds of mass extinctions, species explosions and the acquisition of new degrees of freedom. The human species is now the motor behind one of the biggest species extinctions the earth has ever seen. Should we feel guilty about this? To put our activities into perspective, it is worth realising that if we were not in this position, in all likelihood another intelligent being would eventually evolve and in time face similar dilemmas arising from its own technological inventions. On an optimistic note, we can take comfort from the fact that human activities do not just reduce biodiversity: the human brain itself makes a significant contribution to global biodiversity. And humans will have a major impact in the future through the creation of technical organisms, sowing the seeds of future evolutionary shifts.

As the Red Queen relentlessly goads all organisms into running faster, evolution and biodiversity ensure that new organisms will acquire increasingly complex degrees of freedom at an ever faster pace. In essence, then, the universal utility of biodiversity is the part it plays in the construction of increasingly advanced forms of life.

ACKNOWLEDGEMENTS

The book was written in close cooperation between the author and Frank Veeneklaas, the project manager. Hans-Peter Koelewijn, Rolf Kemmers and Dick Melman made substantive contributions to various drafts of the text. Special thanks go to Menno Schilthuizen for editing the Dutch text and Derek Middleton for his skill, dedication and creativity in translating the original Dutch text into English. This book explores new ground and makes new connections. As the text contains many innovative ideas, we felt the need for independent feedback on the content and lucidity of the text. We therefore sent the first draft to a group of interested people with a broad range of backgrounds. Their suggestions contributed much to the readability and the content of the book. For their valuable contribution we are particularly grateful to all the readers involved from the start of the project: Gert van Maanen, Gert-Jan Hofstede, Herman Eijsackers, Rudy Rabbinge, Jaap Wiertz, Albert Ballast (who suggested using the term 'degrees of freedom'), Jaqueline Mineur, Marnix van Meer, Jennifer Koch, Saskia van den Tweel, Frank Roozen, Joep Dirkx, Peter Schippers, Bert Huisman and Caroline van der Mark.

Printed in the United States
By Bookmasters